Contractor's Exam Book

How to pass the examination
for master builder
and general contractor

John Gladstone
George Perpich
Sandra Perpich

Engineering
Reference
Publications

PAL™ Publications

A division of Direct Brands Inc.
374 Circle of Progress
Pottstown, PA 19464-3800

Tel 1-800-246-2175
Fax 1-800-396-1663

*ISBN : 0-930644-21-2
Library of Congress Catalog Card
number 97-60933*

*Published by Pal Publications
374 Circle of Progress
Pottstown, PA 19464*

First edition published in 1971

PREFACE

The purpose of this Fourth Edition of the **Contractor's Exam Book** is to help candidates successfully prepare for and pass the examination for Certified General, Builder, and Residential Contractor.

Since the appearance of the first edition in 1971, the need for this book has increased tremendously. Many thousands of candidates have applied for contractor licenses in the construction industry as more states and municipalities have written contracting license examinations into their law books. Contractor licensing has itself become a "growth industry." Many colleges and universities now offer degrees in building and construction science. License fees and renewal fees have become a continuing source of revenue. In some states, continuing education credits are required for licensed contractors.

The previous edition of this book was the collaborative effort of three authors: John Gladstone, author of the original first edition as well as eleven other construction industry books and many articles in the technical journals, took his Bachelor's degree at St. Thomas University and his Master's at Vermont College of Norwich University. Formerly a member of the Dade County Construction Trades Qualifying Board, he taught licensing preparation at Miami-Dade Community College. Mr Gladstone presently teaches approved continuing education credits for licensed contractors and journeymen in Miami. George Perpich, past President of the Construction Academy in Dania, Florida was also a member of the Dade County Construction Trades Qualifying Board and held a Florida State Contractor's License as well as a state Mechanical License. He was associated with Gladstone in the construction industry since 1952. Mr Perpich died in 1994 and his name continues to be carried on this title. Sandra Perpich was director of student affairs at the Construction Academy and was active in license preparation work for the past eight years. She attended Florida International University and holds a Certified General Contractor's License for the State of Florida. While some of the original portions remain in this new edition, the current updates for this fourth edition are the sole work of John Gladstone. Any errors that may appear are his, and all suggestions and corrections should be directed to him.

Contents

INTRODUCTION

In the past two decades building construction has fallen under increasingly tight regulation by state, municipal and county authorities. The rationale for this movement has been generally accredited to the need to "protect the public," although according to Michael Pertschuk, chairman of the Federal Trade Commission, "licensing bears little relationship to quality."

Unfortunately, no standards have been set for construction industry licensing examinations. Each municipality, or state usually has its own notion of how to prepare an exam or borrows from some other authority. Although licensing boards throughout the country are usually dominated by local trade practitioners, the laws as well as the license examinations may vary considerably depending upon the experience and integrity of the board members, the experience and intelligence of the political representatives in office, the degree of trade influence imposed by contractor associations and unions, and of course, by established local building practices. A seasoned contractor coming from a major metropolitan area that does not require any license at all may be surprised and chagrined to discover, upon starting a job in some faraway small town, that he or she needs to take an examination for license that will be impossible to pass without extensive preparatory study.

It is not possible to write a book on contractor license preparation that will be useful in every region in the country. Although most exams refer to standard building codes, accepted business practices, and regularly observed construction procedures, there are substantial variations. Obviously some regions need to design and construct for hurricane protection while others are concerned with earthquake protection. Cold-climate exams may stress fire place construction, snow loads on roofs, and hot-water base-board heating systems. In extreme Southern regions exams may include more questions regarding swimming pools, termite protection and air conditioning. Most states have a Mechanics Lien Law; few are identical.

Our approach has been to select one "model" examination that has the greatest commonality with exams in other regions and is at the same time well-balanced and general enough in nature to make it as close to universal as possible. The Florida State General Contractor Examination meets this criterion.

Our preparation method is to make the Candidate as familiar as possible with the type of questions, the areas of concentration, and the reference material he or she will need to study. We assume the reader has a level of education and depth of experience that will enable her or him to understand the terminology and illustrations used.

We have avoided lengthy discourses on building construction techniques and science--that may be found in other books. We have avoided teaching or preaching; there are yet other books for that. By presenting as much exam-related material as possible, we have attempted to fully familiarize the reader with previous exams and working out the solutions to those problems, we feel that the candidate will have enough practice and develop enough self-confidence as to be able to pass the exam handily. We have tried in this volume to write a general treatise on contractor license examination preparation that may be useful either in the formal classroom or as a self-study guide. The Candidate must become acutely aware of the need for disciplined study.

John Gladstone

THE 10 RULES FOR EFFICIENT STUDY.

1. Make a schedule and stick to it. It will raise your level of personal efficiency. It will ease emotional strain and lighten the burden. It will help you master concentration. It will organize your entire family and reduce their interference with your program.

2. Study in the same quiet and well lighted place each time.

3. Keep the top of your desk or study table clear of all unrelated material. Do not wander off your course.

4. Start each study period by the clock, promptly, and end it the same way.

5. If your study sessions are long, take short breaks periodically to relieve tension and stiffness.

6. Study is not reading: As you study, evaluate what you are trying to learn. Why is this expressed this way? What is it for? Can it be done another way?

7. Keep a pencil in your hand while you are studying, and a ruled pad or notebook alongside. A difficult or important passage should be written out; such an expression will help plant the thought firmly in your mind. Summerize ideas in the margin of your books, underscore important passages, rework hard to grasp ideas.

8. Keep a good dictionary of the English language on your desk. Be sure you know the meanings of all the words.

9. Do not get up from your work until it is time. If you need to have a smoke or nibble some pretzels with beer to keep relaxed during your study period, have these things ready before you start work.

10. Review your work constantly. Never, never pass up a review because you feel you have already mastered that lesson.

DURING THE EXAMINATION:

1. Misprints and mistakes may appear on any examination. Raise your hand and question a proctor, but keep working until the proctor arrives. If you think a mistake has been made in the framing of a question, try to make a proper assumption as to what it should have been and proceed to solve the problem. Mark the problem number down and send a letter to the Licensing Board as early as possible following the exam.

2. Do not over-eat during the lunch break. Avoid drowsiness.

3. Use all of the time allowed you. Do not walk out early. When you have finished, review your work completely. Remember, most persons usually fail or pass by only a few points. Every single point is important.

4. Answer every question, do easy ones first, if you have to guess do it by process of elimination, but make sure you mark every question even if it's a guess.

5. Try not to freeze up, if tension is getting you, get up and walk to the restroom.

6. Every question has one or more key words; always look for the "key word" first, it is your main clue.

7. Make sure you read all of the answers before you select one.

8. Keep moving, don't linger on any one question, time is of the essence. Time and accuracy are most important. Make sure you know what you are reading.

9. Again, don't walk out early, take the time to review your work, make sure you colored the right answer on the answer sheet and that you don't have two answers colored.

10. Make notes of all the questions you can remember as soon as you walk out of the exam room.

AFTER THE EXAM:

1. Immediately following the exam, at your earliest possible convenience, make notes while your memory is still fresh. This will be very helpful if you have to retake the exam.

2. For the same reason, go over the problems with some colleagues. If nothing else, it helps relieve the tension while waiting for the results.

3. It's a long wait after the exam--about four weeks before the grades are announced. If you fail, it will probably be by only a few points. Register as early as possible before the next exam, and request a review. There have been many before you have had to sit, two, three, or four times.

OTHER INFORMATION:

Figure 1 on page 23 shows a typical **Answering Sheet** used in most exam rooms. This answer sheet can accommodate a 5-choice answer band or any lesser amount. Exams usually have 4 or 5-choice answer bands.

The **Book List** on page 24 is current for the year 1994, changes may occur with any exam. Page 8 lists convenient sources for current exam books.

All examination books should be properly tabbed. Tabbing systems may usually be purchased through the listed book sources

See the **Appendix** at back of this book for other important information.

APPLICANT
INFORMATION
BOOKLET

CONSTRUCTION EXAMINATION

Instructions for completing your application are contained in this booklet. Please review this information <u>before</u> completing your application.

THIS BOOKLET AND APPLICATION ARE SUBJECT TO ADDITIONS AND/OR REVISIONS.

THE CURRENT REFERENCE LIST(S) ARE VALID THROUGH

April 1997:

INFORMATION BOOKLET OR REVISED REFERENCE LIST YOU MUST WRITE TO:

DEPARTMENT OF BUSINESS & PROFESSIONAL REGULATION CONSTRUCTION INDUSTRY LICENSING BOARD ATTENTION: EXAMINATIONS
7960 Arlington Exp., Ste. 300
Jacksonville, Fl 32211-7467

HOW TO COMPLETE YOUR APPLICATION

Your application for the certification examination must be completed according to the directions of this Applicant Information Booklet, signed, mailed to the address noted on the top of page 1 of the application. All required accompanying documentation must be attached including: photographs; official college transcript or diploma (if necessary); proper verification of your experience; and any other required statements.

Before completing the application, you must:

a) be at least 18 years of age;

b) be of good moral character;

c) possess the necessary experience, properly certified.

You may be prevented from sitting for the examination if your application is not completed properly or is missing proper documentation.

ELIGIBILITY REQUIREMENTS

To be eligible for any portion of the Construction Industry Licensing Board examination, you must show proof of the required experience. It is your responsibility to list your experience in sufficient detail so that it can be determined that you have the proper type of experience. For the categories below, you must meet one of the seven criteria, applicable to you, described in the top portion of the application on page 3 AND meet the specific requirements, as described below for your category on page 4 of the application.

GENERAL CONTRACTOR:

At least one year of *structurally-related experience in commercial construction of four or more stories.

BUILDING CONTRACTOR:

At least one year of *structurally-related experience in commercial, industrial or multiple dwelling residential construction.

RESIDENTIAL CONTRACTOR:

At least one year of *structurally-related experience in residential construction.

*Structurally-related experience is defined as experience in four or more of the following areas: sitework, excavation, footings, piles and pile caps, laying concrete slabs/decks, masonry walls, trusses, wood framing, reinforcement bar, steel erection, column erection and formwork.

SHEET METAL CONTRACTOR:

You must have experience in the manufacture, assembly, erection, installation, dismantling, adjustment, alteration and servicing of ferrous or non-ferrous metal work of U.S. No. 10 gauge or its equivalent or lighter gauge, and of air-handling systems including balancing. You must meet one of the six criteria described in the shaded area on page 3 of the application.

SPECIALTY STRUCTURE CONTRACTOR:

A minimum of one year of * structurally-related experience is required in residential and/or commercial construction using aluminum and allied products, to including fabrication, assembling, handling, erection, installation, dismantling, adjustment, alteration, repair, servicing and design work. You must meet one of the seven criteria, applicable to you,

SPECIALTY STRUCTURE CONTRACTOR *continued*

described in the top portion of the application on page 3.

Structurally-related experience is defined as experience in fou or more of the following areas:
sitework, excavation, foundation, laying concrete (slabs/decks), bloc kneewalls, reinforcement bar and framework.

If you are claiming your experience by virtue of holding a loca aluminum contractor or specialty contractor license, please attach a cop of your certificate of competency.

MECHANICAL CONTRACTOR:

Experience must be in air conditioning; refrigeration; heating boilers; unfired pressure vessel systems; installation of gas, air, vacuum oxygen, nitrous oxide, ink, and chemical line piping; piping an installation of gasoline tanks; pneumatic control piping; condensate piping and piping for lift stations. You must meet one of the seven criteria applicable to you, described in the top portion of the application on pag 3.

CLASS "A" AIR CONDITIONING CONTRACTOR:

At least one year of experience in the installation, maintenanc and repair of systems in excess of 25 tons cooling and 500,000 BT1 heating capacity. Remaining experience must be in the installation maintenance, repair and/or servicing of air conditioning, heating an refrigeration equipment. You must meet one of the seven criteria applicable to you, described in the top portion of the application on pag 3.

CLASS "B" AIR CONDITIONING CONTRACTOR:

At least one year of experience in the installation, maintenanc and repair of heating and cooling systems. Remaining experience must b in the installation, maintenance, repair and/or servicing of ai conditioning, heating and refrigeration equipment. You must meet one o the seven criteria, applicable to you, described in the top portion of th application on page 3.

ROOFING CONTRACTOR:

Experience must be in the installation, maintenance an alteration of all kinds of roofing/waterproofing including but not limited t roll roofing, single-ply, built-up roofing, metal roofing, shingles, shakes tile, modified bitumen, cold process, caulking, sealants and foam. Yo must meet one of the seven criteria, applicable to you, described in the top portion of the application on page 3.

COMMERCIAL POOL CONTRACTOR:

Experience must be in the construction, repair, water treatmen and servicing of swimming pools. At least one year of your experienc must be in the construction of public/commercial pools. You must meet on of the seven criteria, applicable to you, described in the top portion of th application on page 3.

RESIDENTIAL POOL CONTRACTOR:

Experience must be in the construction, repair, water treatmen and servicing of commercial/residential pools. You must meet one of th seven criteria, applicable to you, described in the top portion of th application on page 3.

POOL SERVICING CONTRACTOR:

Experience must be in the servicing, repair, water treatment and maintenance of swimming pools, public or private. You must meet one of the seven criteria, applicable to you, described in the top portion of the application on page 3.

PLUMBING CONTRACTOR:

Experience must be in the installation, maintenance, extension and alteration of all plumbing fixtures and appurtenances in connection with sanitary drainage, venting, public or private water supplies, storm and sanitary sewer lines, water and sewer plants and substations, septic tanks, nitrous oxide piping, fuel oil, gasoline piping, tank and pump installation. You must meet one of the seven criteria, applicable to you, described in the top portion of the application on page 3.

UNDERGROUND UTILITY AND EXCAVATION CONTRACTOR:

Experience must be in the construction, installation and repair on public or private property of main sanitary sewer collection systems, main water distribution systems, and storm sewer collection systems. You must meet one of the seven criteria, applicable to you, described in the top portion of the application on page 3.

SOLAR CONTRACTOR:

Experience must be in the sizing, installation, repair, maintenance, and alteration of domestic portable water heating systems including solar heating panels, solar pool heaters for swimming pools and/or photovolatics systems. You must meet one of the seven criteria, applicable to you, described in the top portion of the application on page 3.

POLLUTANT STORAGE SYSTEMS CONTRACTOR:

Experience must be in the installation, removal, maintenance, extension, and alteration of underground fuel oil, chemical and gasoline pollutant storage tanks and appurtenances, piping and pump installation. You must meet one of the seven criteria, applicable to you, described in the top portion of the application on page 3.

DRYWALL CONTRACTOR:

Experience must be in the installation, maintenance and alteration of all kinds of light steel framing, gypsum board installation and finishing. You must meet one of the seven criteria, applicable to you, described in the top portion of the application on page 3.

QUALIFICATION BY ACTIVE FLORIDA CERTIFIED LICENSE

CERTAIN categories may qualify on the basis of holding a current active certified contractors license for the required number of years, as described below. If you meet the following requirements, check the _first criteria_ listed in the top portion of the application on page 3.

GENERAL CONTRACTOR:

You must possess either an active certified Building Contractor's license or Residential Contractor's license for at least the _previous four years_.

BUILDING CONTRACTOR:

You must possess an active certified Residential Contractor's license for at least the _previous three years_.

QUALIFICATION BY ACTIVE FLORIDA CERTIFIED LICENSE
continued from column 4

COMMERCIAL POOL CONTRACTOR:
You must possess either an active certified Pool Servicing Contractor's license for at least the previous four years or an active certified Residential Pool Contractor's license for at least the previous year.

RESIDENTIAL POOL CONTRACTOR:
You must possess an active certified Pool Servicing Contractor's license for at least the previous three years.

CLASS "A" AIR CONDITIONING CONTRACTOR:
You must possess either an active certified Class "C" Air Conditioning Contractor's license for at least the previous four years or an active certified Class "B" Air Conditioning Contractor's license for at least the previous year.

CLASS "B" AIR CONDITIONING CONTRACTOR:
You must possess an active certified Class "C" Air Conditioning Contractor's license for at least the previous three years.

QUALIFICATION BY ACTIVE FLORIDA REGISTERED LICENSE

If you hold an active "registered" license you may qualify to sit for the "certified" exam category for which you hold the registered license. The registered license must have been active for the previous four years. Check the first criteria listed in the shaded area on page 3 of the application. You MUST submit copies of permits you obtained, one per year, for the past four years as "proof of active licensure".

HOW TO CERTIFY YOUR EXPERIENCE
You must have someone verify your required experience in order to be allowed to sit for the certification examination. Experience must be verified on the "Certification of Experience" section of the application (pages 3 and 4). Duplicate both pages if you need additional forms. DUPLICATION IS NECESSARY IN THE EVENT THAT YOUR EXPERIENCE IS BEING CERTIFIED BY TWO BUILDING OFFICIALS. If you have applied as an "original" applicant in the same category, and that original certification of experience is less than three years old, it is NOT necessary for you to certify your experience again. You must complete pages 1 and 2 only. If the "original" certification of experience is three years old, this office will not retain your file, making it necessary to CERTIFY your experience according to the instructions in this Applicant Information Booklet. Keep a copy of your original application for your records.

The verifier must have "direct knowledge" of your experience. Direct knowledge means the applicant has been in the verifier's employ or the verifier personally knows of the applicant's experience.

If the person verifying your experience is from Florida, he/she must be:

1) a current Florida state certified licensed contractor who may verify ANY category, or

2) a current Florida registered licensed contractor in the category for which you are applying or in a category for which their license allows them to perform or supervise the work in the category for which you are applying, or

Certification of Experience continued from column 5

3) a current licensed architect or engineer, or
4) two building officials employed by a political subdivision of the state who are responsible for inspections of construction improvements.

If your verifier is from outside Florida, he/she must be:
1) a current registered architect or engineer or
2) two building officials employed by a political subdivision of any state, territory or possession of the United States responsible for inspections of construction improvements.

If your verifier is from outside Florida and there is no direct working relationship, experience may be verified by the following:
1) attach a notarized statement listing the projects on which you have worked, <u>AND</u>
2) have the contractor for whom you worked sign the statement verifying that you worked on those projects, <u>AND</u>
3) have an architect, engineer or two building officials certify that the contractor who verified your experience on those projects was the contractor of record for those projects.

If you hold an out-of-state license you may qualify to sit for the exam category in which you meet the requirements by submitting:
1) proof from that state's licensing authority that the out-of-state license was issued as a result of an examination <u>AND</u>
2) supporting documentation from that state's licensing authority describing the scope of work of that license.

<u>IMPORTANT !</u>

All required copies of the application must be signed by you and the person verifying your experience and must contain the verifier's license number and/or seal if applicable. The signature(s) of the verifier's must be notarized. All attachments to the Certification of Experience portion of the application must be signed by the person verifying your experience and must contain the verifier's license number (for a contractor) or seal (for an architect or engineer).

The Contractor Certification Examination is administered on the first and third Wednesday and Thursday of each month. Applications will be processed within ten days after receipt at the Board office. The examination application submitted must be current and be accompanied by the submission of two recent photographs of the applicant taken within 12 months (1 1/2 X 1 1/2 inches in size) and the appropriate fee.

The examination is a two (2) day, open book and multiple choice exam. One part of the exam will test business and finance knowledge and one part will test trade knowledge. An applicant shall be required to retake only the portions of the exam failed. An applicant must pass all portions within three (3) exam attempts after which time all past test scores of the applicant shall be considered invalid and the applicant is required to make an original application and pay all appropriate fees. All three attempts must be completed within a three hundred sixty five (365) day period from the original date of confirmation.

To upgrade a current active certified Division I Residential or Building license, the applicant will not be required to take the business and finance portion of the exam, provided the licensee has not been disciplined, other than a letter of guidance.

Any current certified Division I licensee who has not been disciplined, other than a letter of guidance, may take any Division II examination without being required to take the business and finance portion of the exam.

Any current certified Division II licensee (excluding: Pool Service, Air Conditioning "C" and Asbestos Abatement) desiring to take any other Division II examination, will not be required to take the business and finance portion of the exam provided the licensee has not been disciplined, other than a letter of guidance.

The above provisions apply only to applicants who meet all other requirements and include a copy of their current active license or inactive receipt.

The original application fee for the certification examination administration is $354.00. This fee shall cover the processing of the application and the administration of the examination.

The reexamination fee is three hundred dollars ($300.00). This fee covers the processing of the retake application and the administration of the retake examination.

All application and examination fees are non-refundable and non-transferable.

Any check for payment, returned for any reason by the bank, will be charged a service fee plus the amount of the check. This includes stop payments, bounced checks, etc. If for any reason you believe your payment is wrong, please contact the Board office for further instructions.

Once the application is determined eligible, a confirmation letter will be mailed with instructions to contact the testing firm for scheduling. Space is limited in each site. Candidates MUST make arrangements for testing within thirty (30) days of the date of the confirmation letter and schedule to test within 6 months. Failure to make scheduling arrangements within thirty days will require reapplication. A Candidate Testing Information Booklet will accompany your confirmation letter.

SUBSTITUTING EDUCATION FOR EXPERIENCE

If you possess a construction-related college degree, or any type of college credits at all, they may be applied towards fulfilling your experience requirements. Degrees that are applicable to exam categories are as follows: Civil Engineering & Building Construction are applicable to Division I, Pools & Underground Utility; Architecture is applicable to Division I; Mechanical is applicable to Sheet Metal, Air Conditioning, Solar Water, Mechanical, Plumbing & Underground Utility; and Sanitary Engineering is applicable to Plumbing.

A 4 yr. construction-related degree is equivalent to 3 yrs. experience; a 4 yr. degree in an unrelated field is equivalent to 2 yrs. experience; a 2 yr. construction-related degree is equivalent to 2 yrs. experience; a 2 yr. degree in an unrelated field is equivalent to one year of experience; 1 yr. of accredited college-level courses in the appropriate field of engineering, architecture or building construction is equivalent to 1 yr. of experience. Please note that the columns on the application labeled "Total Man Hours" only need to be completed when (and if) your experience was obtained at the same time that you were attending school (2,000 man hours equals one year of experience).

DEPARTMENT OF BUSINESS AND PROFESSIONAL REGULATION

STAPLE TWO
PASSPORT
SIZED
PHOTOS
HERE
(Print name
on back)

THIS SPACE RESERVED FOR BOARD USE AND
FOR VALIDATION BY REVENUE ONLY

_____ 1st Day Exemption

Mail to:
Construction Industry Licensing Board
c/o 1940 North Monroe Street
Tallahassee, Florida 32399-1006
Attention: Revenue Unit
For Telephone Inquiries: (904) 727-6530

PR20___ 10___ 12___ LI10___ CM40___ REV___
AP10___ AP20___

APPLICATION FOR CERTIFICATION EXAMINATION

FEE: $354.00
THE ABOVE FEE IS NON-TRANSFERRABLE AND NON-REFUNDABLE
MAKE MONEY ORDER, CASHIER'S CHECK OR PERSONAL CHECK PAYABLE TO: DEPARTMENT OF BUSINESS & PROFESSIONAL REGULATION

_____ Original	*FOR BOARD OFFICE USE ONLY*	
Part(s)-To-Take: I II III	Approved By:	Original Date of Confirmation

PART I. EXAMINATION CATEGORIES
CIRCLE EXAM FOR WHICH APPLYING (ONLY ONE)

General	Commercial Pool/Spa	Mechanical	Sheet Metal	Air Conditioning "A"
Building	Residential Pool/Spa	Solar	Roofing	Air Conditioning "B"
Residential	Pool/Spa Servicing	Underground Utility	Plumbing	Pollutant Storage
			Gypsum Drywall	Specialty Structure

PART II. PERSONAL INFORMATION

LEGAL FULL NAME_____
LAST   FIRST   MIDDLE

MAILING ADDRESS_____
STREET   CITY   COUNTY   STATE   ZIP

PERMANENT ADDRESS_____
STREET   CITY   COUNTY   STATE   ZIP

We are required to ask that you furnish the following information as part of your voluntary compliance with Section 2 Uniform Guidelines on Employee Selection Procedure (1978) 43 FR38296 (August 25, 1978.) This information is gathered for statistical and reporting purposes only and does not in any way affect your candidacy for licensure. Note: Under the Federal Privacy Act, disclosure of social security numbers is voluntary. They are requested pursuant to sections 455.203(9), 409.2577, and 409.2598, Florida Statutes, and are used to allow efficient screening of applicants and _____ by a Title IV-D Child Support Agency to assure compliance with child s_____ obligations.

_____/_____/_____ - - -
SOCIAL SECURITY NUMBER

_____/_____/_____ -
DATE OF BIRTH

PLACE OF BIRTH: (City)   (State)   (Nation)

SEX: ____ Male ____ Female   RACE: ____(1)White ____(2)African-American ____(3)Hispanic ____(4)Asian ____(5)Indian ____(6)Other _____

DBPR/CILB/001(REV.8/96)   Page 1

Result of last application: Not Applicable____ Denied____ Failed____ Did Not Show____ Did Not Schedule Within 30 Days____ Did Not Complete Exam Cycle Within 365 Days_____ If you Passed, what category?_____

Attach a copy of all State of Florida certified or registered contractor licenses that you hold or have held, whether revoked, suspended or delinquent.

EDUCATION: CIRCLE LAST YEAR COMPLETED:Grade School 1 2 3 4 5 6 7 8 High School 9 10 11 12
College 1 2 3 4 5 6 7 8   Degree Achieved_____

ATTACH copy of official college transcript, or copy of diploma ONLY when substituting education for experience in accordance with the instructions in the APPLICANT INFORMATION BOOKLET.

PART III.   FINANCIAL RESPONSIBILITY APPLICANT STATEMENT

Please sign below. Any applicant who answers "yes" to any question contained in this Financial Responsibility section of the Application For Certification Examination must supply a complete explanation of the response. The applicant may be required to appear before the Application Review Committee to answer questions regarding such responses.

HAVE YOU (or a partnership in which you were a partner or an authorized representative, or a corporation in which you were an officer or an authorized representative) EVER:

YES NO

☐ ☐ A. Undertaken construction contracts or work that a third party, such as a bonding or surety company completed or made financial settlements?

☐ ☐ B. Had claims or lawsuits filed for unpaid or past due accounts by your creditors as a result of construction operations?

☐ ☐ C. Undertaken construction contracts or work which resulted in liens, suits or judgments being filed?

☐ ☐ D. Had a lien filed against you by the U.S. Internal Revenue Service or Florida Corporate Tax Division? If "yes", you must attach a copy of the Notice of Lien and any payment agreement, satisfaction, Release of Lien or other proof of payment.

☐ ☐ E. Made an assignment of assets in settlement of construction obligations for less than the debts outstanding?

☐ ☐ F. Been charged with or convicted of acting as a contractor without a license, or if licensed as a contractor in this or any state been "subject to" any disciplinary action by a state, county, or municipality? If yes, you must attach a copy of any state, county, municipal or out-of-state disciplinary order or judgment.

☐ ☐ G. Filed for or been discharged in bankruptcy within the past five years? If yes, you must attach a copy of the Discharge Order, Order Confirming Plan or if a Corporate Chapter 7 case, a copy of the Notice of Commencement.

☐ ☐ H. Been convicted or found guilty of, or entering a plea of "nolo contendre" to, regardless of adjudication, a crime in any jurisdiction within the past ten years?

STATEMENT: I affirm the information I have given in this application is true and accurate. I understand any willful falsification constitutes grounds for disqualification. If I am currently a licensee, I understand action may be taken against my license(s) if untrue statements are made in this application. I UNDERSTAND IF I RECEIVE THIS APPLICATION FROM ANY SOURCE OTHER THAN THE CONSTRUCTION INDUSTRY LICENSING BOARD, IT MAY NOT BE COMPLETE.

X_____
APPLICANT'S SIGNATURE   Print Name   Date

_____
Address   City   State   Zip

CERTIFICATION OF EXPERIENCE

PART IV. METHOD USED TO QUALIFY

This form must be completed according to the instructions given in the enclosed APPLICANT INFORMATION BOOKLET.
Refer to the section entitled: ELIGIBILITY REQUIREMENTS. THE PERSON CERTIFYING THE EXPERIENCE MUST HAVE
DIRECT KNOWLEDGE OF THE APPLICANT'S EXPERIENCE. Duplicate this form (front and back) for further certifications.
All attachments pertaining to experience must be signed by verifier and notarized.
NOTE: Insufficient experience is the reason most applications are rejected.

APPLICANT NAME:_____ CATEGORY OF EXAM:_____

ADDRESS:_____

CITY/STATE/ZIP:_____

I AM QUALIFYING FOR THIS EXAMINATION BY: (CHECK ONE)

☐   Holding an active certified or registered_____Florida contractor's license, #_____since_____.

     (CERTAIN categories may qualify on the basis of holding a active certified/registered license. Refer to the
Applicant Information Booklet under the sections entitled QUALIFICATION BY ACTIVE FLORIDA CERTIFIED
    LICENSE and QUALIFICATION BY ACTIVE FLORIDA REGISTERED LICENSE.)

☐   4 year construction-related degree from an accredited college (equivalent to 3 years experience) and 1 year proven
experience (notarized copy of official college transcript, or copy of diploma attached) applicable to the category for
    which you are applying, or

☐   4 year non-construction degree from an accredited college (equivalent to 2 years experience), 1 year experience as
a workman, and 1 year proven experience as a foreman (notarized copy of official college transcript, or copy
    of diploma attached), or

☐   4 years proven experience as a workman or foreman of which at least one year must have been as a foreman, or

☐   2 years proven experience as a workman, 1 year proven experience as a foreman, and 1 year accredited college-level
courses (equivalent to 1 year of experience) in appropriate field of engineering, architecture or building construction
(notarized copy of official college transcript, or copy of diploma attached), or

☐   2 years proven experience as a workman, 1 year proven experience as a foreman, and a 2 year non-construction
degree (equivalent to 1 year of experience, notarized copy of official college transcript, or copy of diploma attached),
    or

☐   2 years proven experience as a workman or foreman of which at least one year must have been as a foreman, and
a 2 year construction related degree (equivalent to 2 years of experience,) applicable to the category for which you
are applying.

TO BE COMPLETED BY PERSON CERTIFYING EXPERIENCE (please print)

I AM ELIGIBLE BASED ON LICENSURE AS A:

_____

Print Name of Person Certifying Experience

____FL current state CERTIFIED CONTRACTOR
____FL current state REGISTERED CONTRACTOR
____Any current REGISTERED ARCHITECT   **Address**
____Any current REGISTERED ENGINEER
____Any current BUILDING OFFICIAL

_____

City/State/Zip

I may be reached by phone for comment, if necessary, at:(_____during business hours.

JOB(S) HELD BY APPLICANT (circle only those applicable)	FROM MONTH/YEAR	TO MONTH/YEAR	*TOTAL HOURS DURING THIS TIME
Workman/Mechanic/Journeyman			
Foreman/Supervisor/Manager/Superintendent			

* SHOW EXPERIENCE IN "HOURS" ONLY WHEN EDUCATION/EXPERIENCE COINCIDE (2,000 hours = one year)

17

TO BE COMPLETED BY APPLICANT

I have read the APPLICANT INFORMATION BOOKLET and reviewed the experience requirements. I understand any false information provided on this form may make me ineligible to take the examination. I also understand that if a license is issued as a result of information I provided on this application and if that information is later reviewed and determined to be incorrect, it could result in the possible loss of license if a license was issued. I certify the foregoing is true and correct.

X_____

NOTARIZED SIGNATURE OF APPLICANT   DATE SIGNED

Before me personally appeared the person named above, known and known to me to be the person described in and who executed the foregoing instrument and acknowledged to and before me that executed said instrument for the purposes therein expressed.

Sworn and subscribed before me this _____ day of _____, 19_____.

X_____
Notary Signature   Notary Seal/Stamp

TO BE COMPLETED BY PERSON CERTIFYING EXPERIENCE

I have read the APPLICANT INFORMATION BOOKLET and reviewed the experience requirements for the above named applicant. I have DIRECT KNOWLEDGE OF THIS APPLICANT'S EXPERIENCE. I UNDERSTAND "DIRECT KNOWLEDGE" DOES NOT MEAN I AM RELYING ON A STATEMENT OR STATEMENTS FROM THE APPLICANT THAT SHE/HE HAS MET THE REQUIREMENT. I FURTHER UNDERSTAND MY LICENSE CAN BE SUBJECT TO DISCIPLINE IF THE INFORMATION GIVEN AND ATTESTED TO BY ME IS FOUND TO BE PURPOSELY MISLEADING AND FRAUDULENT.

X_____

NOTARIZED SIGNATURE OF PERSON CERTIFYING EXPERIENCE   DATE

License #_____

COMPLETE FLORIDA CERTIFIED OR REGISTERED CONTRACTORS LICENSE NUMBER, BUILDING OFFICIAL NUMBER OR REGISTERED ARCHITECT/ENGINEER LICENSE NUMBER (SEAL OR STAMP OF ARCHITECT OR ENGINEER REQUIRED...)

STATE OF   (   )
COUNTY OF   (   )

Before me personally appeared the person certifying experience, named above, known and known to me to be the person described in and who executed the foregoing instrument and acknowledged to and before me that executed said instrument for the purposes therein expressed.

Sworn and subscribed before me this _____ day of _____, 19_____.

X_____
Notary Signature   Notary Stamp/Seal

ALL ATTACHMENTS PERTAINING TO EXPERIENCE MUST ALSO BE SIGNED BY VERIFIER AND NOTARIZED.

18

PART V. VERIFICATION OF EXPERIENCE

LIST AND DESCRIBE WORK PERFORMED DETAILING THE TYPE OF BUILDINGS, PROJECTS AND/OR EQUIPMENT (include the corresponding dates.) Describe your experience in accordance with the "ELIGIBILITY REQUIREMENTS" listed in the APPLICANT INFORMATION BOOKLET for your Licensing Category. ALSO LIST THE COMPANY YOU WORKED FOR OR THE CONTRACTOR(S) THAT SUPERVISED YOUR WORK.
(If applying for the General Contractor Exam, indicate the number of stories with corresponding dates.)

DESCRIBE EXPERIENCE AND WORK PERFORMED	LIST JOBS AND COMPANY AND/OR CONTRACTOR(S) THAT SUPERVISED YOUR WORK	CITY AND STATE WHERE WORK PERFORMED	DATES

19

SIGNATURES OF APPLICANT AND OF PERSON CERTIFYING EXPERIENCE REQUIRED BELOW

TO BE COMPLETED BY APPLICANT
I have read the APPLICANT INFORMATION BOOKLET and reviewed the experience requirements. I understand any false information provided on this form may make me ineligible to take the examination. I also understand that if a license is issued as a result of information I provided on this application and if that information is later reviewed and determined to be incorrect, it could result in the possible loss of license if a license was issued. I certify the foregoing is true and correct.

X_____

NOTARIZED SIGNATURE OF APPLICANT   *DATE*   *SIGNED*

Before me personally appeared the person named above, known and known to me to be the person described in and who executed the foregoing instrument and acknowledged to and before me that executed said instrument for the purposes therein expressed.

Sworn and subscribed before me this _____ day of _____, 19_____.

X_____
Notary Signature   *Notary Stamp/Seal*

TO BE COMPLETED BY PERSON CERTIFYING EXPERIENCE
I have read the APPLICANT INFORMATION BOOKLET and reviewed the experience requirements for the above named applicant. I have DIRECT KNOWLEDGE OF THIS APPLICANT'S EXPERIENCE. I UNDERSTAND "DIRECT KNOWLEDGE" DOES NOT MEAN I AM RELYING ON A STATEMENT OR STATEMENTS FROM THE APPLICANT THAT SHE/HE HAS MET THE REQUIREMENT. I FURTHER UNDERSTAND MY LICENSE CAN BE SUBJECT TO DISCIPLINE IF THE INFORMATION GIVEN AND ATTESTED TO BY ME IS FOUND TO BE PURPOSELY MISLEADING AND FRAUDULENT.

X_____
NOTARIZED SIGNATURE OF PERSON CERTIFYING EXPERIENCE   *DATE*

License #_____
COMPLETE FLORIDA CERTIFIED OR REGISTERED CONTRACTORS LICENSE NUMBER, BUILDING OFFICIAL NUMBER OR REGISTERED ARCHITECT/ENGINEER LICENSE NUMBER (SEAL OR STAMP OF ARCHITECT OR ENGINEER REQUIRED...)
STATE OF   (   )
COUNTY OF   (   )
Before me personally appeared the person certifying experience, named above, known and known to me to be the person described in and who executed the foregoing instrument and acknowledged to and before me that executed said instrument for the purposes therein expressed.

Sworn and subscribed before me this _____ day of _____, 19_____.

X_____
Notary Signature   *Notary Stamp/Seal*

ALL ATTACHMENTS PERTAINING TO EXPERIENCE MUST ALSO BE SIGNED BY VERIFIER AND NOTARIZED.

20

☐ ☐ *Are you requesting special testing accommodations due to any documented disabilities? If so, see section below on SPECIAL TESTING ACCOMMODATION. Be advised that this exam application should not be submitted until you have requested, received and completed an Application For Disability Accommodation.*

SPECIAL TESTING ACCOMMODATION

In accordance with Chapter 61-11.008, Florida Administrative code, the Department will provide special assistance to candidates with documented disabilities. If you have a physical or mental impairment that substantially limits one or more major life activities, you may request special assistance with the examination process.

Please contact the Bureau of Testing immediately to request an Application for Disability Accommodation. The Application for Disability Accommodation must be completed and returned to the Bureau of Testing in accordance with the deadline requirements established for filing your licensure examination application.

RELIGIOUS CONFLICTS

Modification to reporting time or alternate test days may be requested by candidates who, due to their religious beliefs, cannot attend the examination at the scheduled reporting time(s) or on the scheduled date(s). Each request must be made, in writing, by the candidate and sent to the address below. Each request should be accompanied by a letter from the pastor or rabbi from the denomination in which the candidate belongs specifying the religious restrictions that apply. (If the candidate cannot provide this letter, he or she must call the number below.) The request must be received in the Bureau of Testing in accordance with the deadline requirements established for filing the licensure examination application.

<u>Please do not send any correspondence to this address unless it is regarding disabilities or religious conflict</u>

Bureau of Testing
ATTN: Special Testing
1940 North Monroe Street
Tallahassee, Florida 32399-0793
(904) 487-9755

STATE OF FLORIDA
DEPARTMENT OF BUSINESS AND PROFESSIONAL REGULATION
CONSTRUCTION INDUSTRY LICENSING BOARD

GENERAL, BUILDING AND RESIDENTIAL REFERENCE LIST

January 1, 1996 through APRIL 1997

These references are the approved standards issued by the Construction Industry Licensing Board for this examination.

Reference books must remain as published. Photocopies will not be allowed unless written authorization has been granted by the appropriate authorities (publishers) and the Florida Construction Industry Licensing Board. An official letter from the Board must accompany all permitted photo copies.

Beginning February 1, 1996, reference materials with hand-written notes will no longer be allowed in the examination sites. Existing hand-written notes nust be blackened out or whitened out completely prohibiting legibility. Type-written notes also remain prohibited. Materials must remain as published. Underlining in the reference books with pen or highlighter is authorized.

Some of the questions will be based on field experience and knowledge of trade practices.

PLEASE BRING ALL APPROVED REFERENCE MATERIAL TO EACH EXAMINATION SESSION. USE NO OTHER REFERENCES TO ANSWER THE TEST QUESTIONS.

For interpretation of   G C = General Contractor
codes, refer to the   B C = Building Contractor
following: R C = Residential Contractor

****
Reference List

G B R
C C C

X X X   1.   You may select either the reference contained in option A or the one in B below.

           A.   The Guide to Florida's Construction Lien Law, 3rd Edition, as updated 1994, by Neil H. Butler and Louis F. Sisson, III. Profit Guart, Post Office Box 10354, Tallahassee, FL 32302.

           B.   Florida Construction Law Manual, by Larry R. Leiby, 3rd Edition, 1995. Shepard's/McGraw Hill Inc., 555 Middle Creek Pkwy., P.O. Box 350, Colorado Springs, CO 80935.

GC BC RC

VALID FOR EXAMINATIONS ADMINISTERED THROUGH APRIL 1997

G B R
C C C

X X X 2. Contractors Manual, 2nd Edition with 1995 revisions. Association of Builders and Contractors Institute, Inc., 4700 N.W. 2nd Avenue, Boca Raton, FL 33431

X X X 3. Builder's Guide to Accounting, by Michael C. Thomsett, Copyright 1987 or later edition. Craftsman Book Company, 6058 Corte Del Cedro, Carlsbad, CA 92009.

X X X 4. Building Estimators Reference Book, Walker's, 25th Edition. Frank R. Walker Company, P.O. Box 318, 1989 University Lane, Unit "C", Lisle, IL 60532.

X X X 5. Energy Code Excerpts, A Study Guide for the 1993 Florida Energy Efficiency Code for Building Construction, 1995 Edition. State of Florida, Department of Community Affairs, Energy Code Program, 2730 Centerview Drive, Tallahassee, FL 32399. (Note: Chapters 5 & 8 are no longer effective as of 1/1/94)

X X X 6. A. AIA, Document A201, General Conditions of Contract. 1987
X X X   B. AIA, Document A401, Contractor-Subcontractor Agreement. 1987
X X X   C. AIA, Document A701, Instructions to Bidders. 1987
X X X   D. AIA, Document G701, Change Order. 1987
X X X   E. AIA, Document G702, Application and Certificate for Payment, 1992.
X X X   F. AIA, Document G703, Continuation Sheet, 1992.
X X X   G. AIA, Document G706A, Contractor's Affidavit of Release of Liens, 1994. The American Institute of Architects, P.O. Box 60, Williston, VT 05495.

X X X 7. Commentary and Recommendations for Handling, Installing and Bracing Metal Plate Connected Wood Trusses, HIB-91, Truss Plate Institute Inc., 583 D'Onorfio Drive, Suite 200, Madison WI 53719

X X X 8. Principles and Practices of Heavy Construction, Fourth Edition, Copyright 1993, by Ronald C. Smith and Cameron K. Andres. Prentice-Hall Publishers, 200 Old Tappan, Old Tappan, NJ 07675

X X X 9. Design and Control of Concrete Mixtures, Thirteenth Edition, 1988 with 1994 revisions, by Steven H. Kosmatka and William C. Panarese. Portland Cement Association, 5420 Old Orchard Road, Skokie, IL 60077-1083.

X X X 10. Placing Reinforcing Bars, Sixth Edition, CRSI Concrete Reinforcing Steel Institute, 933 North Plum Grove Road, Schaumburg, IL 60173.

X X 11. Formwork For Concrete, (sp-4) Fifth Edition, 1989 or later edition. American Concrete Institute, P.O. Box 19150, 22400 W. Seven Mile Rd., Detroit, MI 48219.

X X X 12. Recommended Specifications for the Application and Finishing of Gypsum Board, GA-216-93 or later edition. Gypsum Association, 810 First Street N.E., #510, Washington, DC 20002.

GC BC RC

VALID FOR EXAMINATIONS ADMINISTERED THROUGH APRIL 1997

G B R
C C C

X X X   13.   Standard Building Code, 1994 Edition. Southern Building Code Congress International, Inc. 900 Montclair Road, Birmingham, AL 35213-1206.

X X X   14.   Blueprint Reading for the Building Trades, Copyright 1985. Craftsman Book Company, 6058 Corte Del Cedro, Carlsbad, CA 92009.

X X X   15.   Code of Federal Regulations, (OSHA) 29 Part 1926, including Subpart P, Revised as of July 1st, 1994, or later edition. The Government Printing Office Bookstore, 100 West Bay Street, Jacksonville, FL 32202.

(END OF LIST)

GC BC RC

VALID FOR EXAMINATIONS ADMINISTERED THROUGH APRIL 1997

WHERE TO GET THE BOOKS

Few areas in the United States have a technical bookstore. To further complicate matters, most technical bookstores do not stock the full line of books required for nation-wide examinations. In many cases it is necessary to purchase the required books by mail or phone. Some book stores do specialize in exam books; it is best to go to one or two sources for your books. The following is a list of some bookstores known to stock and ship books required for the exams including local and national code books. One of the most complete exam book suppliers is Construction Bookstore Inc. They have several outlets around the country with headquarters in Gainesville, Florida and may be reached at 1-800: 253 0541.

Builder's Book Depot
1033 East Jefferson, Suite 500
Phoenix, AZ 85034   602: 252-4050

Construction Bookstore Inc.
1830 NE Second Street
Gainesville, FL 32602   1-800: 253-0541

Downtown Book Center Inc.
247 SE First Street
Miami, FL 33131   305: 377-9939

Engineer's Bookstore
748 Marietta St.
Atlanta, GA 30318   404: 892 1669

Florida Exam Bookstore
7650 Pembroke Road
Hollywood, Fl 33023   305: 620-2277

Irvine Sci-Tech Bookstore
4040 Barranca Parkway
Irvine, CA 92714

Opamp Technical Books
1033 Sycamore Ave.
Los Angeles, CA 90038   213: 464-4322

TAB YOUR BOOKS

One of the most important strategies to beat the test writers is a fool-proof system of tabbing and highlighting all the books that will be required in the examination room. The authors of this book have developed such a system. The complete instructions for tabbing and highlighting the reference books are prepared and updated for every exam. Several of the listed book sources carry current editions of GUIDE FOR HIGHLIGHTING AND TABBING: Florida State License Examination. Following are instructions for preparing tabs. An excellent tabbing system is also offered by Construction Bookstore, see page 11

1. From your local stationery/office supply store, purchase the required number of ALCO SURE STICK INDEX TABBING, size 1/3", stock number 10381. Order one box of red, one box of yellow, and as many boxes of clear as required. One box equals approximately 75 tabs.

2. Type or letter the index label as per instructions for individual books. Use three lines of label for each legend as indicated below, and

type each legend twice, directly beneath one another.

3. Tear off strip, allowing three lines, and fold the blank line inside, allowing the typed legend to appear on both sides as shown below.

2nd line, back ——→ △ HEATING ——
↑ —— 1st line, inside   3rd line, front

This three thickness fold is necessary to keep the index label from slipping out of the Tab.

4. Insert folded label into Tab and cut Tab to fit label.

5. Review book tabbing instructions for each book and make a pencil mark on each page (right-hand) to receive a Tab Index, keeping a distance of about 3/8" between the vertical dimensions of each Tab.

6. Line up the Tab on the exact place on the page--making sure you are holding only one page--bend the skirt back sharply to remove backing paper from inside. Press skirt to sheet firmly.

   Clear Tabs are used for all text copy, red tabs for indexes and contents and yellow Tabs for tables and special references.

The importance or index tabbing of the reference materials used in the exam room cannot be over stressed. Every second gained during the examination is a valuable step towards passing. The index system used here has been carefully designed to provide the utmost workability under extreme time pressure.

S T U D Y   S C H E D U L E

This is your work plan; give it your meticulous attention. Develope a realistic schedule of the most possible hours you can devote to this program--then stick to it with an iron discipline! Draw an X through hours you intend to set aside for study under each day and post this schedule in a visible place for constant reference.

	SUNDAY	MONDAY	TUESDAY	WEDNESDAY	THURSDAY	FRIDAY	SATURDAY	
7:00								7:00
8:00								8:00
9:00								9:00
10:00								10:00
11:00								11:00
12:00								12:00
1:00								1:00
2:00								2:00
3:00								3:00
4:00								4:00
5:00								5:00
6:00								6:00
7:00								7:00
8:00								8:00
9:00								9:00
10:00								10:00
11:00								11:00

FORM LETTER

To receive your application for the Florida State examination, send this form letter. Fill in the proper blanks stating whether you wish to take the General Contractor, Building Contractor, Residential Contractor category, the date of the examination and the Southern, Northern or Central part of the state.

State of Florida
Department of Professional Regulations
Florida Construction Industry
Licensing Board
7960 Arlington Expressway, Suite 300
Jacksonville, Florida 32211-7467

Dear Sir:

Please forward all necessary information and application to

the State examination, on_____for the category

of_____Contractor. I request that I take the

examination in the_____part of the State.

    NAME_____

    ADDRESS_____

    CITY_____STATE_____ZIP_____

    TELEPHONE NUMBER_____

- -

A similar letter may be drafted for any state or municipality whose laws provide for contractor exams. Licensing boards will usually respond to a phone call as well as a letter, but it is wise to send a letter and keep a record of all correspondence and photo copies of all applications, documentation, etc., in a permanent file.

Note: Phone number for the CILB is 904/359-6310

BASIC MATH FOR
TAKING PROFESSIONAL EXAMINATIONS

Mathematics as required for the Florida State Certified Contractors License is not a matter of having knowledge of higher mathematics such as calculus and statistics, but rather a good understanding of the function of numbers.

Mathematics is the language of numbers; therefore, to have a proper understanding of the language, the first step is to understand the symbols associated with the numbers and how these symbols dictate direction of the numbers. The add (+) symbol and the subtract (-) symbol are the symbols we deal with most frequently along with the multiplication (x) symbol and the division (÷) symbol. The add (+) and subtract (-) symbols are used in the addition and subtraction process and also in the multiplication (x) and division (÷) process.

In the addition and subtraction process the answer is always governed by the sign of the greater number or product of numbers. If no sign is shown it is assumed that the number is (+).

$$\text{EXAMPLE:} \quad 20 + 5 = 25$$
$$20 - 5 = 15$$
$$-20 + 5 = -15$$

Associating the above with a scale having a zero axis;

| -20 | -15 | -10 | -5 | 0 | +5 | +10 | +15 | +20 |

For example a thermometer reading +20°C and rising 5°C would read 25°C (20 + 5 = 25). If the temperature fell 5°C it would read 15°C (20 - 5 = 15). In the same respect if the reading was -20°C and the temperature rose 5°C it would read -15°C (-20 + 5 = -15). In the same respect if the reading was -20°C and the temperature rose 5°C it would read (-20 + 5 = - 15).

A mathematical background will certainly be helpful in any contractors license examination, but a good exposure of elementary mathematical concepts and operations is all that is really necessary for the Florida State Certified Contractors examination.

It is assumed that the candidate is already familiar with the four fundamental operations of arithmetic mentioned above.

Once the candidate has mastered the use of the calculator and refreshed himself or herself with the basics, it is wise to do practice exercises.

Whenever examples appear in this book, they should be carefully studied and followed step-by-step using the calculator. Simply reading them will prove fruitless.

THE PORTABLE ELECTRONIC CALCULATOR

The calculator is the most important item required for the examination because the largest part of the exam requires the manipulation of numbers.

A quiet battery operated calculator without read-out tape is required in the exam room. It is recommended you have a 12-digit calculator with a large keyboard and a readout window tilted upward for easy viewing. The unit must have a percent key (%) and a square root key (√‾‾‾). The reason for the 12-digit calculator is that the eight digit units will zero out when large numbers are computed and the number of digits in the summation are greater than the calculator can handle.

The calculator used for this discussion is a Sharp VX-2134, 12-digit, which employs the memory key process. Should you have a unit similar to the Texas Instrument Constant Memory series the functions are slightly different.

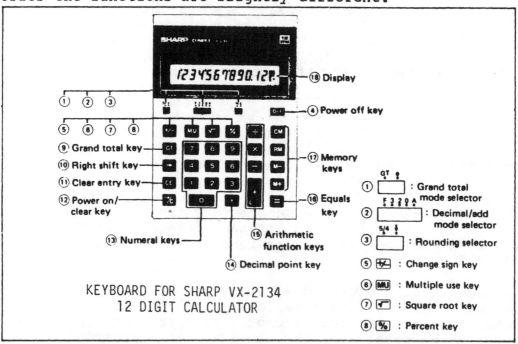

KEYBOARD FOR SHARP VX-2134
12 DIGIT CALCULATOR

It is also recommended you have more than one calculator. Many candidates have gone into the exam room with only one calculator. When it failed due to the wearing out of batteries or malfunction, they were at a complete loss. There is no way the problems in the exam can be solved within the allotted time limit without the use of a calculator. It is necessary to have both calculators with the same key board functions.

Each calculator manufacturer will adopt a slightly different style of keyboard. Some calculators become a bit elaborate, which is alright if you are experienced with the calculator's functions and are able to use them. If the operation of the calculator is unfamiliar to you, do not purchase a calculator with more information than has been described. A booklet of instructions is supplied with each new calculator. Be sure to read the booklet thoroughly.

When reading the booklet (again referring to the memory key type calculator) be aware of the DECIMAL/ADD MODE SELECTOR AND THE (5/4) ROUNDING SELECTOR. The DECIMAL/ADD MODE SELECTOR locates the placing of the decimal point; the (5/4) ROUNDING SELECTOR raises or lowers the decimal equivalent past the 0.5 mark. Each calculator manufacturer may vary the design of these two functions. It is most important to read the instruction booklet for the proper setting of these two functions.

The calculator does not have a read-out tape; therefore, accuracy is more important than speed. Check each number in the read-out window before proceeding. Insert the next number, making sure the number is correct and the decimal is in the proper place.

ADDITION AND SUBTRACTION

When adding or subtracting, the plus ⊕ or minus ⊖ key must be pressed before the entry of the desired number. Numbers can be added or subtracted in any sequence and the calculator will carry the sign of the larger number.

> EXAMPLE: adding; 100 ⊕ 250.43 ⊕ 92.3 ⊜ 442.73
> (when there is no sign the number is
> considered plus ⊕.)
>
> adding; 100 ⊕ ⊖ 250.43 ⊕ ⊖ 92.3 ⊜
> -242.73 (the calculator will display the minus
> (-) sign.)

MULTIPLICATION

When multiplying, the functions must be entered into the calculator as they are written out. Each time a whole number or mixed number is entered into the calculator the times key ⊗ must be pressed. When the product is required press the equal key ⊜.

> EXAMPLE: 2 ⊗ 2 ⊜ 4
> 2 ⊗ 2 ⊗ 2 ⊜ 8
> 2.45 ⊗ 3.68 ⊗ 4 ⊜ 36.064

DIVISION

The division function is similar to the multiplication function; the numbers must be placed into the calculator as written. The top number of the division process (numerator) must be placed in the calculator first then the division key ⊟ is pressed. Then the bottom number of the division process (denominator) is placed into the calculator and the equal ⊟ sign key is pressed.

        EXAMPLE: 2 ⊟ 4 ⊟ .5
                  4 ⊟ 2 ⊟ 2

If the numerator is larger than the denominator the product will be a whole number if the denominator is larger than the numerator the product will be a decimal part of the whole number.

A series of numbers may also be divided.

        EXAMPLE: 400 ⊟ 2 ⊟ 4 ⊟ 2 ⊟ 25

A combination of multiplication and division may be performed.

        EXAMPLE: 4 ⊠ 8 ⊟ 4 ⊟ 8

COMPLEX DIVISION

Complex division may be performed on the calculator but the division process has no memory storage.

        EXAMPLE: Calculate the following,

$$\frac{\pi \times (0.96 + 5.66 - 4.032)}{1.6 \times 9}$$ - one atmosphere, psi (14.7)

First perform the functions within the brackets;
(0.96 ⊞ 5.66 ⊟ 4.032)

Then, ⊠ 3.14 to complete the numerator above the line.

Now, ⊟ 1.6 ⊟ 9 ⊟ 14.7 ⊟ -14.136; the display window will read (14.136-).

Notice, the single multiplication below the line (denominator) becomes a double division. Once the numerator has been calculated all multiplying signs in the denominator change to dividing signs. But additions and subtractions must be mentally or manually stored.

EXAMPLE: Solve for cfm in the following equation:

$$\frac{360,000 \text{ Btuh}}{.075 \times .24 \times 60 \text{ min} \times (40F - 10F)}$$

= 360,000 ÷ 30 ÷ .075 ÷ .24 ÷ 60 = 11,111 cfm

When doing this example, the 36,000 was first entered. Then the
△t resulting from (40F - 10F = 30) and entered as a division
function below the numerator. Do not attempt to mix-multiply and
divide values below the line (denominator), take every x sign as
÷ sign.

FRACTIONS/DECIMALS

When dealing with fractions always convert them to their decimal
equivalent. To convert a fraction to a decimal equivalent,
divide the numerator by the denominator;

$$\frac{3, \text{ numerator}}{4, \text{ denominator}} = .75, \text{ quotient}$$

If the fraction is an improper fraction (a whole number plus a
fraction, 2-3/4) convert only the fractional part (as above) then
include it with the whole number i.e., 2.75.

3 ÷ 4 = .75   2 + 3/4 = 2 + .75 = 2.75

When converting feet and inches into feet and a fractional part
of a foot, divide the inch part only as if it were a fraction,
using 12" as the denominator.

EXAMPLE: Convert 4'-6" to feet   6" = 6/12 of a foot
6" ÷ 12" = .5; therefore, 4'-6" = 4.5'

To convert feet plus a fractional part of a foot, multiply
the fraction part of the foot by 12 inches.

EXAMPLE: Convert 4.5' to feet inches.
.5 ⊗ 12"/ft = 6"; therefore, 4.5' = 4'-6"

DECIMAL/ADD MODE SELECTOR AND THE 5/4 ROUNDING SELECTOR

When using decimal equivalents, the DECIMAL/ADD MODE SELECTOR
(F320A) and the 5/4 ROUNDING SELECTOR (5/4↓) become optional.
The purpose of these keys is to pre-set the number of decimal
places. In the "F" position (F320A) the decimal will float

out to the maximum number of places in the 12-digit display
window. The "3" position reduces the answer to 3 decimal
places. The "2" position reduces the answer to 2 decimal places,
and the "0" position deletes the decimal places allowing only the
whole number to appear in the display window.

32

EXAMPLE: Using the DECIMAL/ADD MODE SELECTOR find the
          decimal equivalent for 1/3.

    a. (F320A) 1 ÷ 3 = 0.33333333333
        ↑

    b. (F320A) 1 ÷ 3 = 0.333
         ↑

    c. (F320A) 1 ÷ 3 = 0.33
          ↑

    d. (F320A) 1 ÷ 3 = 0.
           ↑

EXAMPLE: Using the (5/4) ROUNDING SELECTOR.
          Find the sum of 1.6432 + 1.5321 + 1.4239

    a. (F320A) (5/4) 1.6432 + 1.5321 + 1.4239
        ↑   ↑   = 4.5992

    b. (F320A) (5/4) 1.6432 + 1.5321 + 1.4239 = 4.59
        ↑   ↑

    c. (F320A) (5/4) (This is the rounding off position)
        ↑   ↑ 1.6432 + 1.5321 + 1.4239 = 4.60

For most calculations it is recommended that position "3"
(F320A) and (5/4) be used, this will display an answer with

a 3-place decimal, rounded off.

If the fourth digit after the decimal is less than 5 do not
round off the third digit, but if the fourth digit after the
decimal is 5 or greater, raise the third digit by one. The
(5/4) ROUNDING SELECTOR will do this automatically.

When multiplying a whole number by a decimal, the product is
always a lesser number.

        20 (x) .5 = 10, product

When dividing a decimal by a whole number the quotient will
always be smaller.

        .5 ÷ 20 = .025, quotient

When dividing a whole number by a decimal, the quotient will
always be larger.

        20 ÷ .5 = 40, quotient

OTHER CALCULATORS

The Sharp Compet VX-2134 12-digit calculator used in the above discussion is representative of available calculators at most outlets for this kind of merchandise. The choices are many and the prices vary considerably. The Sharp Compet with 12-digit display has 28 function keys. As a comparison, the illustration below shows a typical 25 key, 8-digit display window equal to a Unisonic, Sharp EL-8131 or other competitive, smaller models.

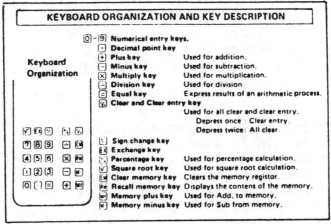

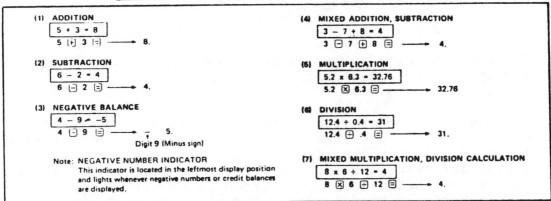

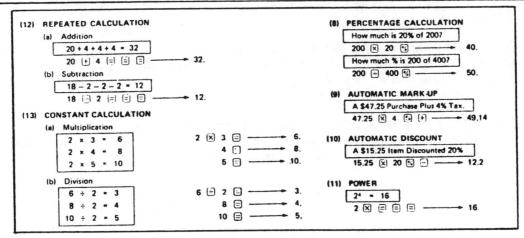

THE RULE OF SIGNS

Words must, of course, have their grammar to give them precise meanings; take, for example, the word "jump". Everyone understands the meaning of jump, but to give it precise or specific meaning we need to know that it is a verb and can be used in a past tense sentence, "The cow jumped over the moon" a present tense sentence, "The cow is jumping over the moon", or a future tense sentence, "the cow will jump over the moon"

In a somewhat similar manner mathematics has a grammar known as signed numbers to indicate the specific function, or operation of any number. The signed numbers are + (positive) and –(negative). To complete the "sentence structure" of numbers they may have function symbols such as – (minus), x (times), $\sqrt{\quad\quad}$ (radical), and be phrased into an equation such as,

$$2 \times 3 + 4 = 10 \quad \text{or} \quad A = \pi R^2 \quad \text{or} \quad 2 + 2 = 4$$

These equations require simple multiplication or addition. An equation is simply a statement of equality, i.e., something is equal to something else. An equation may also be a complex statement of equality. It is only required to read and express simple equations in the examination room. Equations are usually interpreted by the equal sign (=) indicating equality,

$$12 - 4 = 8 \quad \text{and} \quad 12 \text{ in.} = 144 \text{ sq inches}$$

In common usage a number is considered positive if it is not accompanied by a plus sign (+); therefore, the number 12 really indicates + 12.

The common calendar used in the United States (Gregorian calendar) designates the year of this book as 1985 meaning literally, 1985 years after Christ. The years preceeding Christ are stated BC meaning before Christ. Using signed numbers and deleting the reference to Christ the present year would correctly read + 1985. And 1369 years BC would be – 1369.

When calculating an event that happened 1369 years BC the operation of signed numbers would be necessary. The calculation would read 1985 - (-1369) = + 3354, meaning the event would now be 3354 years old.

THE RULE OF SIGNS for the above statement is: A minus sign before a parenthesis changes the sign of every term inside when the parenthesis is removed; thus,

$$a - (-b) = a + b \quad \text{or} \quad 8 - (-4) = 8 + 4 = 12$$
$$a - (+b) = a - b \quad \text{or} \quad 8 - (+4) = 8 - 4 = 4$$

--See **RULE OF SIGNS** next page --

In mathematical equations, the unknown symbol χ is usually entered by typewriter or non-script face looking like this, "x" or "X". In either of these cases it could be confused or mistaken for the unknown symbol χ. Therefore, the (times) x symbol may be substituted by a centrally placed period,

$$2 \cdot 4 = 8$$

or the numbers may be placed in brackets,

$$(2)(4) = 8$$

Whenever two or more numbers are used in an associated phrase without any symbol separating them, the operation indicated is multiplication. Note, for example, the equation,

$$C = \pi d$$

which states that the circumference of a circle is equal to 3.14 multiplied by its diameter. No symbol is used to designate times (x); the multiplication symbol is assumed.

Brackets are frequently used to signal multiplication, for example, 5 Btuh ÷ (.075)(.24)(60) $\triangle$t = cfm, which means the cfm is equal to the sensible heat,

$$Btuh \div .075 \text{ x } .24 \text{ x } 60 \text{ x } (T_2 - T_1).$$

RULE OF SIGNS

To Multiply Numbers With Like Signs

Rule: Multiply and place a plus before the product.

EXAMPLE: $(-5) \text{ x } (-6) = +30$ $(+5) \text{ x } (+6) = +30$

To Multiply Numbers with Unlike Signs

Rule: Multiply and place a minus sign before the product.

EXAMPLE: $(+5) \text{ x } (-6) = -30$ $(-10) \text{ x } (+5) = -50$

To Divide Numbers With Like Signs

Rule: Divide and place a plus sign before the quotient.

EXAMPLE: $(+12) \div (+6) = +2$ $(-12) \div (-6) = +2$

To Divide Numbers With Unlike Signs

Rule: Divide and place a minus sign before the quotient.

EXAMPLE: $(+12) \div (-6) = -2$ $(-24) \div (+3) = -8$

The above may be reduced to the RULE OF SIGNS for multiplication;

1. | + x + = + |

2. | - x - = + |

3. | + x - = - |

4. | - x + = - |

The calculator will change signs with the | +/- | key, although in the beginning it is recommended that the operator use the rule of signs manually. The + and - keys cannot be used in the multiplication and division operations.

EXAMPLE: $\dfrac{40 \times (-20) \times 10}{(10) \times (-5)}$ = ?

40 (x) 20 (x) 10 (=) 8000, manual minus = -8000
        10 (x) 5 (=)   50, manual minus = -50

8000 (÷) 50 (=) 160, manual plus = +160

∴ $\dfrac{40 \times (-20) \times 10}{(10) \ (-5)}$ = +160

To Add Numbers With Like Signs

Rule: Add the numbers, regardless of the sign, then give the like sign,

EXAMPLE:
```
  +7        -7        -163
  +4        -4        - 89
  +11       -11       -252
```

The calculator will perform this operation,

(-) 163 (-) 89 (=)   display, (252-)

To Add Numbers With Unlike Signs

Rule: Separate the positives from the negatives and form two groups. Then subtract the smaller group from the larger group and give the sign of the larger group.

EXAMPLE: 25 + 96 + 64 -29 -33 -44 =

Manual Separation   **Calculator**

```
+25    -29     185      25 (+) 96 (+) 64 (-) 29 (-) 33 (-) 44
 96     33    -106
 64     44    + 79              (=) display, (79)
+185   -106
```

EXAMPLE: 25 - 29 + 64 - 33 - 44 + 10 =

<u>Manual</u> <u>Separation</u>   <u>Calculator</u>

```
+25      -29      -106      25 ⊖ 29 ⊕ 64 ⊖ 33 ⊖ 44 ⊕ 10
 64       33   (-)+ 99
 10       44    = -  7            (=) display, (7-)
+99     -106
```

<u>Subtracting</u> <u>Signed</u> <u>Numbers</u>

Signed numbers may be subtracted with the use of the RULE OF
SIGNS. In the examination room, the candidate may encounter
problems dealing with excavations below El. 0.00, transit
problems, or low temperature conversion problems of Fahrenheit to Centigrade. Any of these may require the application
of the RULE OF SIGNS. The use of the rule may be mental/
manual manipulation, or through the use of the +/- key. When
using the +/- key make sure the right sign ends up in the
final answer. This comes with practice.

When subtracting signed numbers, the RULE OF SIGNS may be
applied by simply changing the minus sign of every term to a
plus sign whenever a minus sign (-) appears before a
parenthesis. Thus, a - (-b), by changing the term becomes a
+b (Rule 4)

        EXAMPLE: 2 - (-4) =
                  2 ⊕ 2 ⊜ 4

In this illustration, the operator applies Rule 4, (minus x
minus = plus) to change the sign mentally and then plugs the
problem into the calculator to get a read-out of 4.

        EXAMPLE: In digging a footer the existing
                   elevation has been determined to be
                   +0.83' and the plans indicate that
                   the bottom of the footer is to be
                   established at elevation -1.65'.
                   How deep must the builder dig down?

                   +0.83 - (-1.65) = +2.48   (rule 2)

Applying the RULE OF SIGNS:
        a -(-b) = a + b, = 0.83 ⊕ 1.65 ⊜ 2.48

        EXAMPLE: Convert -40° Fahrenheit to Celsius.

            Step 1. From the conversion formula,
                     °C =(°F - 32) x 5/9

                     (-40) - (+32)
                     (-40) + (-32) = -72

Step 2. $-72 \times 5/9 =$
          $5/9 = 5$ ⊘ 9 ⊜ .555
          $=$ (-72) ⊗ .555 ⊜ $-40°C$, (rule 4)

EXAMPLE: An engineers survey shows that in the construction of a 28 story building the bottom of the base footing must be at elevation -14.0'. An old three story building must be removed and after demolition the existing uniform grade is at elevation -3.0'. How deep must the builder dig to establish the bottom of the new base footer?

           $-3 -(-14) = 11$ ft. down   (rule 2)

EXAMPLE: Subtract, $20-5-(-18)-(-3) =$ changing the signs within the parenthesis mentally/manually, the problem reads,

           $20 - 5 + 18 + 3 =$
           20 ⊖ 5 ⊕ 18 ⊕ 3 ⊜ 36

           Whenever the plus ⊕ or minus ⊖ precedes parenthesis, the RULE OF SIGNS must be applied.

THE PERCENT KEY

The percent key (%) is a short step for avoiding the use of a decimal to find the percent of a number. Applying this concept to solving a simple problem of sales tax,

           EXAMPLE: Find the amount of sales tax at 5% on a transaction of $269.60.

Conventional; $5\% = 5$ ⊘ 100 ⊜ .05, decimal

                .05 ⊗ 269.60 ⊜ $13.48, display
                                      (14 movements)

Percent Key;   269.60 ⊗ 5 ⊛ display, $13.48
                                      (9 movements)

SQUARE ROOT KEY

To find the square root of a number on the calculator, plug in the number and press the (√‾‾) key. Read the answer in the display window.

EXAMPLE: 1. Find the square root of; 25

25 (√‾) display, (5)

2. Solve for 46

46 (√‾) display, (6.7823299)

To find the square root of a fraction, convert the fraction to a decimal and follow the operation above,

EXAMPLE: Solve for 3/4

3 (÷) 4 (=) .75

.75 (√‾‾) display, (.866)

Standard calculators do not have a cube root $\sqrt[3]{}$ key. If a table of powers and roots is not handy the following rule will help to extract the $\sqrt[3]{}$ of a number with the use of a standard calculator: Enter the number for which the cube root is desired, enter the (√‾‾) key twice, enter (x), enter the correct multiplier, read approximate $\sqrt[3]{}$.

Number (n)	Multiplier
1-25	1.30
26-40	1.35
41-60	1.40
61-75	1.45
76-100	1.50

EXAMPLE: Extract the $\sqrt[3]{60}$

60 (√‾) (√‾) (x) 1.40 (=) 3.896

From the Table of Powers and Roots the actual cube root of 60 = 3.915

MEMORY FUNCTIONS

The calculator has four memory keys for the storage of figures which may be used in future calculations.

Memory Clear(CM, MC, MT etc.) Memory Recall (RM, MR, MS etc.)

Memory Minus (M-)   Memory Plus (M+)

The Memory function is used to store a series of numbers that have been calculated by the process of addition, subtraction, multiplication, or division into Memory + (M+) or Memory - (M-). Another operation may then be computed and again stored in Memory until the product is to be recalled. Only the summation or product of a computation may be stored in Memory. When all Memory calculations are completed, the Memory Clear key (CM) will disengage memory and clear.

EXAMPLE: Compute cost of following merchandise with discount,

5 units at $500.00   3 units at $200.00   2 units at $600.00

Total discount 10% and total sales tax 5%

STEP	ENTER KEY	PRESS KEY	ENTER KEY	PRESS KEY	WINDOW DISPLAY	ENTER KEY
1.	5	(X)	500	(=)	2500	(M+)
2.	3	(X)	200	(=)	600	(M+)
3.	2	(X)	600	(=)	1200	(M+)
4.		(RM)			4300	
5.		(X)	10	(%)	430	(M-)
6.		(RM)			3870	
7.		(X)	5	(%)	193.50	(M+)
8.		(RM)	⟶	ANSWER =	4063.50	
9.		(CM)	Memory Clear			

MATH QUIZ NO. 1

ADD:
1. 251.05 + 33.582 + 6953.01 + 52 + .03 =

2. 953.115 + 66.05 - 57.32 + 63.1 - 362.132 =

3. -2 + (-8) + (-6) =

4. -8 + 6+ (-7) + 4 + 16 + (-2) =

MULTIPLY:
5. 90 X 3 X 5.27 =

6. (-90) x (-3) =

7. (-90) x (+3) =

8. .245 x .78 x 1.35 =

9. (-30) x (-8) x (+4) x (-20) =

DIVIDE:
10. $\frac{120}{4}$ =

11. $\frac{-426}{4}$ =

12. $\frac{520 \times 123}{43560}$ =

13. $\frac{.333 \times 49 \times 87}{27}$ =

14. $\frac{\pi \times 90 \times 50}{37 \times 10}$ =

15. $\frac{(8 + 27) \times 60 \times .0175}{6 \times 27 \times .10}$ =

CHANGE THE FOLLOWING TO FEET:

16. 2' 4"= _____ 25' 3"= _____ 11"= _____ 5' 1"= _____

MULTIPLY AND GIVE ANSWER IN FEET:

17. 6' 9" x 112' 2" =

18. 16' 3" x 8' 6" x 3' 4" =

ANSWER SHEET

MATH QUIZ NO. 1

1. 251.05 ⊕ 33.582 ⊕ 6953.01 ⊕ 52 ⊕ .03 ⊜ 7289.672

2. 953.115 ⊕ 66.05 ⊖ 57.32 ⊕ 63.1 ⊖ 362.132 ⊜ 662.813

3. ⊖ 2 ⊕ ⊖ 8 ⊕ ⊖ 6 ⊜ -16

4. ⊖ 8 ⊕ 6 ⊕ ⊖ 7 ⊕ 4 ⊕ 16 ⊕ ⊖ 2 ⊜ 9

5. 90 ⊗ 3 ⊗ 5.27 ⊜ 1422.9

6. -90 ⊗ -3 ⊜ +270

7. -90 ⊗ +3 ⊜ -270

8. .245 ⊗ .78 ⊗ 1.35 ⊜ .257985

9. 30 ⊗ 8 ⊗ 4 ⊗ 20 ⊜ -19,200

10. 120 ⊘ 4 ⊜ 30

11. -426 ⊘ 4 ⊜ -106.5

12. 520 ⊗ 123 ⊘ 43560 ⊜ 1.468

13. .333 ⊗ 49 ⊗ 87 ⊘ 27 ⊜ 52.577

14. 3.14 ⊗ 90 ⊗ 50 ⊘ 37 ⊘ 10 ⊜ 38.189

15. 8 ⊕ 27 ⊗ 60 ⊗ .0175 ⊘ 6 ⊘ 27 ⊖ .10 ⊜ 2.269

16. 4 ⊘ 12 ⊕ 2 ⊜ 2.333

    3 ⊘ 12 ⊕ 25 ⊜ 25.250

    11 ⊘ 12 ⊜ .916

    1 ⊘ 12 ⊕ 5 ⊜ 5.083

17. 9" ⊘ 12" ⊕ 6' ⊜ 6.750'
    2" ⊘ 12" ⊕ 112' ⊜ 112.167'
    6.750' x 112.167' ⊜ 757.127 ft

18. 3" ⊘ 12" ⊕ 16 ⊜ 16.250
    6" ⊘ 12" ⊜ .5 ⊕ 8 ⊜ 8.500
    4" ⊘ 12" ⊕ 3 ⊜ 3.333
    16.250 ⊗ 8.500 ⊗ 3.333 ⊜ 460.371 ft

FRACTIONS AND DECIMALS

The fraction and decimal are a part of an integer (whole number). For the purpose of construction math it is recommended that all fractions be converted to their decimal equivalents.

A fraction is always written over a division sign, 1/2, 9/16, 15/11. The top half of the fraction is known as the numerator and the bottom half is known as the denominator. Fractions are either proper fractions or improper fractions. Proper fractions will always have a value less than one (smaller than a whole number). Improper fractions will always have a value greater than one (a whole number plus its fractional part). The proper fraction always has a numerator less than the denominator (1/2, 3/4, 7/8) wherein the improper fraction always has a numerator greater than the denominator (13/4, 5/2, 18/5).

To convert a fraction to a decimal divide the numerator into the denominator. (1/2 = 0.5)

If the fraction is an improper fraction it will have an integer (whole number) and its decimal equivalent (13/2 = 6.5).

A mixed number is a whole number and a fraction (3 1/3). To convert the decimal equivalent use only the fractional part of the number (1/3 = .333) then include the whole number 3.333.

Most important is to keep the fraction in its right perspective. When converting inches to the fraction of a foot make sure it is inches and not a fraction of an inch.

EXAMPLE: 12'-6" = 12.5'

       12'-6 3/8"   (first the fraction of an inch must be converted to inches.)

       3/8" = .375" = 6.375"   then,

       6.375/12 = .531' = 12.531'

Whenever you wish to convert a decimal equivalent back to a fraction multiply the decimal portion only of the total number by a chosen denominator and obtain the numerator.

EXAMPLE: 12.531' using only .531 x 12 = 6.375"

       .375 x 12 = 8 = 3/8' = 12' 6 3/8"

When using any type of conversion be sure that the numbers
are in the same category;

EXAMPLE:

Cubic feet per hour, and not cubic feet per minute.
Pounds per square inch, and not pounds per square foot.
Cubic feet and not cubic yards.
Temperature centigrade, and not temperature farenheit.

(you must always change the terms to the same meaning)

ADDITION AND SUBTRACTION

When adding or subtracting numbers that include a decimal
place the decimals must always be in a vertical line.

EXAMPLE:

ADD	1.360	SUBTRACT	− 4.323
	.003		− 5.552
	.254		−16.820
	16.302		− .003
	17.919		−26.698

Note: The calculator will perform this process and place the
decimal in its proper place.

MULTIPLICATION

When multiplying numbers with a decimal the product must have
the amount of decimals that are in the numbers that are being
multiplied. Count the decimals starting from right to left.

EXAMPLE:

3.14	126.3	12.532
2.5	.003	.002
1570	.3789	.025064
628		6 places
7.850	4 places	
3 places		

Note: The calculator will perform this process and place the
decimal in its proper place.

DIVISION

When dividing a whole number by a decimal, move the decimal
in the denominator and the numerator until the denominator is
a whole number.

EXAMPLE: 70 divided by .05

$$.05 \overline{)70.00} \qquad = \qquad 5 \overline{)7000}$$

When dividing a decimal by a whole number, the decimal in the
answer must be placed directly above the numerator.

EXAMPLE: .25 divided by 20

$$
\begin{array}{r}
.0125 \\
20 \overline{)\, .25} \\
20 \\
50 \\
40 \\
100 \\
100
\end{array}
$$

When a decimal is divided by a whole number the quotient will
always be a smaller number; when dividing a whole number by
a decimal the result will be greater.

EXAMPLE: $\dfrac{.5}{20}$ = .025

$\dfrac{20}{.5}$ = 40

Note: The calculator will perform this process and place the
decimal in its proper place.

ALGEBRA AND EQUATION SOLVING

Algebra is simply a continuation of the language of numbers. The expression 2 + 2 = 4 is an algebraic equation. The expressions of algebraic symbols and formulae usually denote and replace the arithmetic process with letters in place of numbers. Algebra is simply 2 + 2 = 4.

An algebraic equation is always separated by the equal (=) sign. This indicates that the left side of the equation must have the same value as the right side of the equation. Again referring to the equation 2 + 2 = 4 note that both sides of the equation have the value of 4. To keep the equation in balance it must be understood that a number may be moved to the other side of the equal sign although when doing so the characteristic of that number must be changed to keep the equation in balance.

The equation 2 + 2 = 4 may also be written 2 x 2 = 4, with the two simple numbers:

$$2 + 2 = 4$$
$$2 \times 2 = 4$$

The rules of algebra must be established:

In the first equation:

$$2 + 2 = 4$$

by applying the **RULE OF SIGNS** for addition and subtraction, numbers may be moved from one side of the equal sign to the other providing the sign is changed. Removing a 2 from the left side of the equation to the right side the equation will read;

$$2 = 4 - 2$$

therefore, 2 = 2. Moving the 4 to the left side of the equation it will read;

$$2 + 2 - 4 = 0$$

therefore, 0 = 0.

The relationship of 0 = 0 can be applied only to the add and subtract process. It cannot be applied to the multiplication and division process.

In the second equation;

$$2 \times 2 = 4$$

which is the multiplication process. The multiplication

nomenclature must first be understood which is written as
follows;

$$2 \quad 2 = 4$$
$$2 \times 2 = 4$$
$$(2)(2) = 4$$
$$2(2) = 4$$

also in algebra when replacing numbers with letters
multiplication nomenclature is written as follows;

$$ab = c$$
$$a \times b = c$$
$$(a)(b) = c$$
$$a(b) = c$$

Portions of the algebraic equation may be placed in
parenthesis ( ) or brackets [ ], which indicates that
quantities are grouped together and considered a single
quantity.

EXAMPLE: $[2 \times 6 (4 + 8)] = 144$

When a problem is shown as in the example above the inner
parenthesis (or brackets) must first be calculated then
continue to calculate outward.

When a number is next to a parenthesis or bracket with no
symbol between them the process is to multiply.

EXAMPLE: $6(4 + 8) = 72$

Either the inner parenthesis may be computed first $(4 + 8) =$
12 then multiply $6 (12) = 72$, or multiply into the
parenthesis $6(4 + 8) = 6 \times 4 = 24$ and $6 \times 8 = 48$, then $24 +$
$48 = 72$.

Referring back to the equation;

$$2 \times 2 = 4$$

numbers may be moved to the opposite side of the equation by
reducing the equation with a common denominator;

$$\frac{2 \times 2}{2} = \frac{4}{2} \qquad 2 = 2$$

or by transposing a number to the opposite side of the
equation by making it the denominator.

$$2 \times 2 = 4 \qquad 2 = \frac{4}{2} \qquad 2 = 2$$

What cannot be done in multiplication is to remove a total
quantity from one side of an equation making that side equal
to (0). To remove a total quantity from one side of an
equation both sides of the equation must be divided by the
sum so that one side of the equation will be equal to one.

$$2 \times 2 = 4 \qquad \frac{2 \times 2}{4} = \frac{4}{4} \qquad 1 = 1$$

The prime function of the principles of algebra is to solve
equations with unknown factors. For purposes of the Florida
State Certified Contractors Exam it is only necessary to
solve for one unknown. Using X as the unknown, again
referring to the equation;

$$2 + 2 = 4$$

replacing one quantity with X the equation will read;

$$X + 2 = 4 \qquad X = 4 - 2 \qquad X = 2$$

also in the second equation;

$$2 \times 2 = 4$$

replacing one quantity with X the equation will read;

$$X \times 2 = 4 \qquad X = \frac{4}{2} \qquad X = 2$$

wherein if $\qquad X \times 4 = 4 \qquad X = \frac{4}{4} \qquad X = 1$

in the same equation if

$$X \times 4 = 0 \qquad X = 0$$

for any number multiplied or divided by 0 = 0.

The equation $\frac{4}{2} = 2$ is the division process. Replacing 4 with
X the equation reads $\frac{X}{2} = 2$. To solve for X let the product
of the known side become the numerator and make the
denominator equal to (1).

$$\frac{X}{2} = \frac{2}{1} \quad \text{then cross multiply} \quad \frac{X}{2} \diagdown \diagup \frac{2}{1}$$

$$X = 2 \times 2 \qquad X = 4$$

With one unknown, quantities may be added, subtracted, multiplied, or divided,

EXAMPLE: (addition)

$$X + 5 = 10 \qquad X = 10 - 5 \qquad X = 5$$

Proof: $5 + 5 = 10$

EXAMPLE: (subtraction)

$$X - 10 = 20 \qquad X = 20 + 10 \qquad X = 30$$

Proof: $30 - 10 = 20$

EXAMPLE: (multiplication)

$$X \times 5 = 20 \qquad X = \frac{20}{5} \qquad X = 4$$

Proof: $4 \times 5 = 20$

EXAMPLE: (division)

$$\frac{X}{5} = 25 \qquad \frac{X}{5} = \frac{25}{1} \qquad X = 25 \times 5 \qquad X + 125$$

Proof: $\frac{125}{5} = 25$

To expand on the equation:

$$[\, 2 \times 6\, (4 + 8)\,] = 144$$

Replacing $(4 + 8)$ with $(X + 8)$

and solving for X

$$[\, 2 \times 6\, (X + 8)\,] = 144$$

multiplying

$$(2 \times 6) = 12$$

$$[\, 12\, (X + 8)\,] = 144$$

$$12X + 96 = 144$$

$$12X = 144 - 96$$

$$12X = 48$$

$$X = \frac{48}{12} \qquad X = 4$$

Replacing 2 x 6 with X x 6

and solving for X

$$[X \times 6 (4 + 8)] = 144$$

$$4 + 8 = 12$$

$$X \times 6 \times 12 = 144$$

$$72 X = 144$$

$$X = \frac{144}{72}$$

$$X = 2$$

The language of algebra also involves superscripts and subscripts. The superscript X^n indicates a number raised to a power, which means a number multiplied by itself (see page 39) any number of times as directed by the superscript.

$$2^2 = 2 \times 2 = 4$$

$$2^3 = 2 \times 2 \times 2 = 8$$

$$2^4 = 2 \times 2 \times 2 \times 2 = 16$$

Should the superscript read X^{-n} the indication is that it is the reciprocal of X^n or $\frac{1}{X}$.

$$2^{-2} = \frac{1}{2^2} = \frac{1}{4} = .25$$

$$2^{-3} = \frac{1}{2^3} = \frac{1}{8} = .125$$

The subscript X_n has no significant mathematical meaning except to define the condition of the number.

EXAMPLE:

cfm_1	cfm_2	rpm_1	rpm_2
$cfm_1 = 3000$	$cfm_2 = 2000$	$rpm_1 = 500$	$rpm_2 = 333.3$

One of the functions of algebra is to solve equations or mathematical statements. The above nomenclature, cfm and rpm are associated with the equation;

$$\frac{cfm_2}{cfm_1} = \frac{rpm_2}{rpm_1}$$ indicates that cfm varies directly with the fan speed

Solving for rpm_2 and using the above numbers;

$$\frac{2000}{3000} = \frac{rpm_2}{500}$$ cross multiplying $\frac{2000}{3000} = \frac{rpm_2}{500}$

$$2000 \times rpm_2 = 2000 \times 500$$

$$rpm_2 = \frac{1,000,000}{3000} = 333.3 \ rpm_2$$

RATIOS AND PROPORTIONS

RATIOS

A ratio is a relationship of two or more numbers to a quantity, or a fractional part of a whole quantity. A ratio may be written as 1/4 or 1:4. The numbers 1 and 4 are called terms.

A ratio of 1:4 signifies that one item is four times as large as the other item.

> EXAMPLE: If a pully on a belt driven motor has a ratio of 1:4 and the driver pully is 6" in diameter, how large is the driven pully?

$$1:4 \qquad as \qquad 6:X$$

A ratio of 1:4 signifies that the driven pully is four times as large as the driver pully; therefore,

   X is four times larger than 6; 4 x 6 = 24" driven pully.

If the ratio were 2:5 and a 6" driver pully given, the ratio signifies that one item is three times larger than the other; therefore, multiply both terms of the ratio by 3:

$$2 \times 3 = 6 \qquad and \qquad 5 \times 3 = 15$$

equaling a ratio of 6:15, or a 15" driven pully.

> EXAMPLE: If a brine water mixture had a ratio of 2:3, how many gallons of brine and how many gallons of water are in 2000 gallons of mixture.

> SOLUTION: With a ratio of 2:3 the total quantity would be 5 (2 + 3); therefore, 2/5 of the total is brine and 3/5 of the total is water.

   Brine = 2/5 = .40 or 40% = .40 x 2000 = 800 gal.
   Water = 3/5 = .60 or 60% = .60 x 2000 = 1200 gal.

Regardless what the ratios are, the terms can always be expressed as a fractional part of a whole or in percentages of 100%.

A ratio of 8:12:20 is a percent of 40 (8 + 12 + 20).

8/40	.20	or	20%
12/40	.30	or	30%
20/40	.50	or	50%
	1.00		100%

PROPORTIONS

A proportion is a relationship between two ratios.

$$\text{ratio (1)} \quad 2:4$$
$$\text{ratio (2)} \quad 4:8$$

$$\text{Proportion} \quad 2:4 \; :: \; 4:8$$

The numbers 2 and 8 are known as the extremes and the numbers 4 and 4 are known as the means. The product of the extremes must be equal to the product of the means.

Proportions may also be written as:

$$\frac{2}{4} = \frac{4}{8} \quad \text{and by cross multiplying} \quad \frac{2}{4} \diagdown \frac{4}{8}$$

$2 \times 8 = 16$ and $4 \times 4 = 16$; therefore $16 = 16$

and the same results are accomplished.

Replacing one known with X, solving for X

$$X:4 :: 4:8$$
$$8X = 16$$
$$X = 2$$

EXAMPLE: What is the length of the missing dimension?

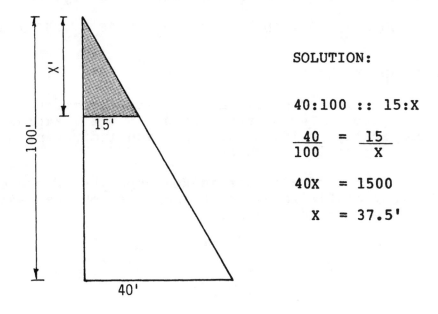

SOLUTION:

$$40:100 :: 15:X$$

$$\frac{40}{100} = \frac{15}{X}$$

$$40X = 1500$$

$$X = 37.5'$$

SQUARES, CUBES AND POWERS

The square of any number is that number multiplied by itself once, it is called the second power.

$$\text{The square of } 4 = 4 \times 4 = 16$$

Expressed algebraically; $4^2 = 16$. The small number ^{2} is called a superscript or an exponent. It is placed to the right of, and above any number.

$$6^2 = 6 \times 6 = 36, \text{ or 6 squared} = 36.$$

The exponent means that the number (in this case six) shall be used as a factor twice. If the exponent is 6^3 it means that the number six shall be used as a factor three times; 6 x 6 x 6 = 216. We may say that six raised to the third power is 216, or

$$\text{The cube of } 6 = 6 \times 6 \times 6 = 216$$

A number may be raised to any power. Often the exponent is used as a shorthand to avoid the use of many zeros. For example thirty-six million may be written 36,000,000. Notice that this number contains six zeros. It may be written algebraically, 36×10^6 which simply signals the reader to add six zeroes to the number 36. 144,000,000,000 may be expressed as 144×10^9 .

Conversely, when a superscript is written with a minus sign in front of it as 10^{-3} it signals the reader to move three decimals out. So $10^{-3} = 0.001$, and $36 \times 10^{-3} = 0.036$, and $36 \times 10^{-2} = 0.36$.

The use of superscripts are very important for converting metric SI units. For example, to convert fpm to m/s (feet per minute to meters per second) the conversion factor is 5.080×10^{-3} . Therefore to convert 900 fpm to m/s,

$$900 \times 5.080 \times 10^{-3} = 900 \times .00508 = 4.57 \text{ m/s}$$

ROOTS AND RADICALS

The root of a number is the opposite operation of raising it to the power. The square root of 16 = 4. Expressed algebraically; $\sqrt{16} = 4$. The sign indicating square root is $\sqrt{}$, called a radical.

The cube root of a number is the opposite of the cube. The cube root of 216 is 6, it is indicated by the radical $\sqrt[3]{}$; therefore, it can be stated;

$$\sqrt[3]{216} = 6$$

CONSTRUCTION CALCULATORS

A number of handy construction calculators are available from several sources at prices ranging from $40 to $90. These hand-held calculators are excellent in the exam room as well as in the field for manipulating common mathematical formulas, stair calculation, board feet plus dollar cost, rafter calculations, etc. Most proctors will allow these calculators. Here are some sample calculations using a "Construction Master" or Measure Master-Plus."

COMMON RAFTER CALCULATIONS (PITCH KNOWN)

The roof you are working on has a 7:12 pitch (7 inches). You know the overall span of the building is 23 feet. To what length should you cut the common rafters?

NOTE: In all rafter calculations you still need to allow for the ridge adjustment and eave overhang.

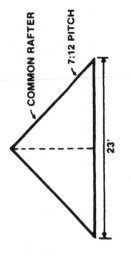

COMMON RAFTER

7:12 PITCH

23'

- Press PITCH to enter known pitch

  `7.` INCHES

- Press RUN to enter known run

  `11 - 6.` FEET INCHES

- Press SLOPE

- **Displays rafter length**

  `13 - 3 ⁴⁹/₆₄` FEET INCHES

  Elapsed time: 19 seconds

TOTAL BOARD FEET PLUS DOLLAR COST

You are buying 100 2 x 4 x 8's, at a cost of $213 per thousand board feet. How many board feet are you buying, and what is your cost?

8"

4"

2"

- Press BOARD FEET key for Board Feet Mode

- Enter dimensions for 100 2 x 4 x 8's

- Press BOARD FEET

- **Displays total board feet**

  `533.` BDFT

- Enter unit price

- Press UNIT PRICE

- **Displays total cost of 2 x 4 x 8's**

  `213.` BDFT

  `113.5290` BDFT

  Elapsed time: 23 seconds

STAIR CALCULATION — STRINGER LENGTH

Find the stringer length if the floor-to-floor rise is 8' 10⅝" and the total run of the stairway is 10' 10". The risers are 7⁹/₃₂" high.

STRINGER RULE: For stringer calculation, the rise of the stairway equals the floor-to-floor rise minus the length of the last riser.

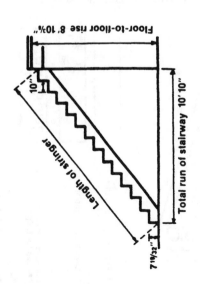

Floor-to-floor rise 8' 10⅝"

Total run of stairway 10' 10"

Length of stringer

10"

7 ¹⁹/₃₂"

- Enter known rise

  `8 - 10. ³/₈` FEET INCHES

- Subtract riser height

  `7 ¹⁹/₃₂` FEET

- Press RISE to enter as rise of stairway

- Enter total run; press RUN to enter

  `10 - 10.` FEET

- Press SLOPE

- **Displays stringer length**

  `13 - 7 ¹¹/₆₄` FEET INCHES

  Elapsed time: 32 seconds

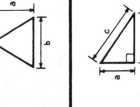

NOTE π = 3.14159

Area of circle
πr^2

Volume of a sphere
$\dfrac{4\pi r^3}{3}$

Volume of a cylinder
$\pi r^2 h$

Area of an ellipse
πab

Area of a triangle
$\dfrac{ab}{2}$

Length of a hypotenuse
$c = \sqrt{a^2 + b^2}$

NOTE: You can bypass this formula by using the RUN, RISE and SLOPE keys.

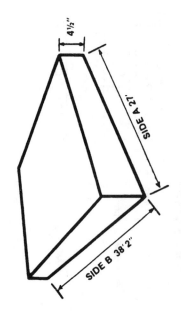

SIDE A 27'
SIDE B 38' 2"
4½"

VOLUME CALCULATION
LANDSCAPING TOP SOIL OR CONCRETE SLAB

You are going to fill an area 38'2" long, 27" wide and 4½" deep. First calculate the total area, then determine the total cubic yards required.

- Enter total length of side A

  `27.` FEET

- Multiply by length of side B

  `38-2.` FEET

- Displays square feet of area

  `1030.5` SQ FEET

- Multiply by depth of fill

  `4. ½` INCHES

- Displays volume (cubic feet)

  `386.4375` CU FEET

- Press CONVERT TO

- Press YARDS

- **Displays volume (cubic yards)**

  `14.3125` CU YARDS

Elapsed time: 26 seconds

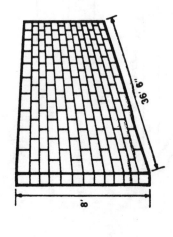

36' 6"
8'

MASONRY — ESTIMATING BRICKS

How many standard bricks (2½" x 3¾" x 8") with ½" joints would be required for a wall measuring 36'6" long and 8' high?

- Enter height of wall

  `8.` FEET

- Divide by height of brick and joint

  `2. ¾` INCHES

- **Displays number of bricks needed for height**

  `34.90909`

- Enter into MEMORY

- Enter width of wall

  `36-6.` FEET

- Divide by length of brick and joint

  `8. ½` INCHES

- **Displays number of bricks needed for width**

  `51.52941`

- Multiply by number of bricks in height (in memory)

- **Displays total bricks needed**

  `1798.845`

Elapsed time: 54 seconds

SI UNITS - METRICS QUIZ

1. 14 psi pressure = _____

    A. 96.5 kPa   C. 96,500 Pa
    B. 96.5 kJ/kg   D. A and C

2. 50 ft^3 = _____

    A. 141.5 m^3 C. 14.15 m^3
    B. .01415 m^3 D. 1.415 m^3

3. 2650 ft. = _____

    A. 808 m^3/s   C. 808 mm
    B. 808 m/s   D. 808 m

4. 15.4 psi = _____

    A. 1.06 kPa   C. 10.6 kPa
    B. 1060 kPa   D.   106 kPa

5. 52 psf = _____

    A. 35,850 kN/m   C. 2490 Pa
    B. 358,540 Pa   D. 358.5 kPa

6. 100 board ft. = _____ m^3

7. .5 miles = _____ kilometers

8. 6.4 acres = _____ Hectares

9. 9 gallons = _____ liters

10. 4 sq. miles = _____ km^2

ANSWER SHEET

SI UNITS - METRIC QUIZ

1. (A) 14 psi pressure X 6.89 = 96.46 kPa

2. (D) 50 cu ft X 2.832 X 10^{-2} = 1.416 m^3

3. (D) 2650 ft X 3.048 X 10^{-1} = 807.72 m

4. (D) 15.4 psi X 6.89 = 106.11 kPa

5. (C) 52 psf X 47.9 = 2491 Pa

6. 100 bd ft = 100 sq ft X 1/12'(thick) = 8.3 cu ft X .0283
    = .234 m^3

7. .5 miles X 1.61 = 0.805 kilometers

8. 6.4 acres X 4.047 X 10^{-1} = 2.59 hectares

9. 9 gallons X 3.79 = 34.11 liters

10. 4 sq miles X 2.59 = 10.36 km^2

MENSURATION

$$\left[\begin{array}{c} A = \text{area}, \ r = \text{radius}, \ d = \text{diameter}, \ c = \text{circumference} \\ h = \text{height}, \ b = \text{base} \end{array} \right]$$

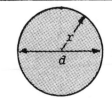

CIRCLE

Area $= \pi r^2 = 3.14 \times r \times r$

Circumference $= 2\pi r = 3.14 \times d$

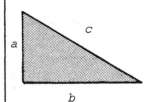

RIGHT-ANGLED TRIANGLE

Area $= \dfrac{ab}{2}$

Length of a side $=$

$c = \sqrt{a^2 + b^2}$ $c = (a \times a) + (b \times b)$

$a = \sqrt{c^2 - b^2}$ $a = (c \times c) - (b \times b)$

$b = \sqrt{c^2 - a^2}$ $b = (c \times c) - (a \times a)$

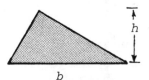

ACUTE-ANGLED TRIANGLE

Area $= \dfrac{bh}{2} = b \times h / 2 = $ base $\times$ height$/2$

CIRCLE

The shaded area (ring)

$A = \pi (R^2 - r^2) = 3.14 \times (R + r) \times (R - r)$

ARC OF A SQUARE

The shaded area (fillet or spandrel)

$A = 0.215r^2 = 0.215 \times r \times r$

HEXAGON

Area $= 2.6 \ S^2$ (actually 2.598)

$2.6 \times S \times S$

OCTAGON

Area $= 4.83s^2$ (actually 4.828) $= A = 4.83 \times S \times S$

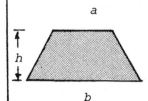

TRAPEZOID

Area $= \dfrac{(a + b)\, h}{2}$

height $\times$ a + b/2

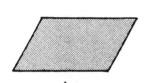

PARALLELOGRAM

Area $= hb$
    height $\times$ base

$h = A/b$

$b = A/c$

TABLE 1

CIRCUMFERENCE FOR 90° OF A CIRCLE OF ANY GIVEN RADIUS

For other than 90° use the formula Circumference = 0.01745 x radius x degrees

RADIUS	90°	RADIUS	90°	RADIUS	90°
1	1.5	36	56.5	71	111.5
2	3.125	37	58.125	72	113.06
3	4.75	38	59.75	73	114.625
4	6.25	39	61.25	74	116.187
5	7.875	40	62.75	75	117.375
6	9.5	41	64.375	76	119.375
7	11.0	42	66.0	77	121.0
8	12.5	43	67.5	78	122.5
9	14.125	44	69.125	79	124.06
10	15.75	45	70.75	80	125.625
11	17.25	46	72.25	81	127.187
12	18.875	47	73.75	82	128.75
13	20.5	48	75.375	83	130.375
14	22.0	49	77.0	84	131.875
15	23.5	50	78.5	85	133.5
16	25.125	51	80.06	86	135.06
17	26.75	52	81.75	87	136.625
18	28.25	53	83.25	88	138.25
19	29.875	54	84.75	89	139.75
20	31.5	55	86.75	90	141.375
21	33.0	56	88.0	91	143.0
22	34.5	57	89.5	92	144.5
23	36.125	58	91.06	93	146.06
24	37.75	59	92.75	94	147.75
25	39.25	60	94.25	95	149.25
26	40.875	61	95.75	96	150.75
27	42.5	62	97.375	97	152.375
28	44.0	63	99.0	98	154.0
29	45.5	64	100.5	99	155.5
30	47.125	65	102.06	100	157.06
31	48.75	66	103.625		
32	50.25	67	105.312		
33	51.75	68	106.75		
34	53.375	69	108.375		
35	55.0	70	110.0		

TABLE TO CONVERT INCHES TO A DECIMAL OF A FOOT

Inches	0	1	2	3	4	5	6	7	8	9	10	11
		.0833	.1667	.2500	.3333	.4167	.5000	.5833	.6667	.7500	.8333	.9167
1/16	.0052	.0885	.1719	.2552	.3385	.4219	.5052	.5885	.6719	.7552	.8385	.9219
1/8	.0104	.0937	.1771	.2604	.3437	.4271	.5104	.5937	.6771	.7604	.8437	.9271
3/16	.0156	.0989	.1823	.2656	.3489	.4323	.5156	.5989	.6823	.7656	.8489	.9323
1/4	.0208	.1042	.1875	.2708	.3542	.4375	.5208	.6042	.6875	.7708	.8542	.9375
5/16	.0260	.1094	.1927	.2760	.3594	.4427	.5260	.6094	.6927	.7760	.8594	.9427
3/8	.0312	.1146	.1979	.2812	.3646	.4479	.5312	.6146	.6979	.7812	.8646	.9479
7/16	.0365	.1198	.2031	.2865	.3698	.4531	.5365	.6198	.7031	.7865	.8698	.9531
1/2	.0417	.1250	.2083	.2917	.3750	.4583	.5417	.6250	.7083	.7917	.8750	.9583
9/16	.0469	.1302	.2135	.2969	.3802	.4635	.5469	.6302	.7135	.7969	.8802	.9635
5/8	.0521	.1354	.2188	.3021	.3854	.4688	.5521	.6354	.7188	.8021	.8854	.9688
11/16	.0573	.1406	.2240	.3073	.3906	.4740	.5573	.6406	.7240	.8073	.8906	.9740
3/4	.0635	.1458	.2292	.3125	.3958	.4792	.5625	.6458	.7292	.8125	.8958	.9792
13/16	.0677	.1510	.2344	.3177	.4010	.4844	.5677	.6510	.7344	.8177	.9010	.9844
7/8	.0729	.1562	.2396	.3229	.4062	.4896	.5729	.6562	.7396	.8229	.9062	.9896
15/16	.0781	.1615	.2448	.3281	.4115	.4948	.5781	.6615	.7448	.8281	.9115	.9948

EXAMPLE: Convert 2½" to decimal of a foot. Enter first line at 2 inches. Drop down to horizontal equivalent for ½ in. Read answer = .2083 ft.

Convert 9½ inches to decimal of a foot. Enter first line at 9 inches. Drop down to horizontal equivalent for ¼ inch. Read answer = .7708 ft.

POWERS AND ROOTS

n	n²	n³	√n	∛n
1	1	1	1.0000000	1.0000000
2	4	8	1.4142136	1.2599210
3	9	27	1.7320508	1.4422496
4	16	64	2.0000000	1.5874011
5	25	125	2.2360680	1.7099759
6	36	216	2.4494897	1.8171206
7	49	343	2.6457513	1.9129312
8	64	512	2.8284271	2.0000000
9	81	729	3.0000000	2.0800838
10	100	1000	3.1622777	2.1544347
11	121	1331	3.3166248	2.2239801
12	144	1728	3.4641016	2.2894285
13	169	2197	3.6055513	2.3513347
14	196	2744	3.7416574	2.4101423
15	225	3375	3.8729833	2.4662121
16	256	4096	4.0000000	2.5198421
17	289	4913	4.1231056	2.5712816
18	324	5832	4.2426407	2.6207414
19	361	6859	4.3588989	2.6684016
20	400	8000	4.4721360	2.7144176
21	441	9261	4.5825757	2.7589242
22	484	10648	4.6904158	2.8020393
23	529	12167	4.7958315	2.8438670
24	576	13824	4.8989795	2.8844991
25	625	15625	5.0000000	2.9240177
26	676	17576	5.0990195	2.9624961
27	729	19683	5.1961524	3.0000000
28	784	21952	5.2915026	3.0365890
29	841	24389	5.3851648	3.0723168
30	900	27000	5.4772256	3.1072325
31	961	29791	5.5677644	3.1413807
32	1024	32768	5.6568542	3.1748021
33	1089	35937	5.7445626	3.2075343
34	1156	39304	5.8309519	3.2396118
35	1225	42875	5.9160798	3.2710663

n	n²	n³	√n	∛n
36	1296	46656	6.0000000	3.3019272
37	1369	50653	6.0827625	3.3322219
38	1444	54872	6.1644140	3.3619754
39	1521	59319	6.2449980	3.3912114
40	1600	64000	6.3245553	3.4199519
41	1681	68921	6.4031242	3.4482172
42	1764	74088	6.4807407	3.4760266
43	1849	79507	6.5574385	3.5033991
44	1936	85184	6.6332496	3.5303483
45	2025	91125	6.7082039	3.5568933
46	2116	97336	6.7823300	3.5830479
47	2209	103823	6.8556546	3.6088261
48	2304	110592	6.9282032	3.6342412
49	2401	117649	7.0000000	3.6593057
50	2500	125000	7.0710678	3.6840315
51	2601	132651	7.1414284	3.7084298
52	2704	140608	7.2111026	3.7325112
53	2809	148877	7.2801099	3.7562858
54	2916	157464	7.3484692	3.7797631
55	3025	166375	7.4161985	3.8029525
56	3136	175616	7.4833148	3.8258624
57	3249	185193	7.5498344	3.8485011
58	3364	195112	7.6157731	3.8708766
59	3481	205379	7.6811457	3.8929964
60	3600	216000	7.7459667	3.9148676
61	3721	226981	7.8102497	3.9364972
62	3844	238328	7.8740079	3.9578916
63	3969	250047	7.9372539	3.9790572
64	4096	262144	8.0000000	4.0000000
65	4225	274625	8.0622577	4.0207258
66	4356	287496	8.1240384	4.0412400
67	4489	300763	8.1853528	4.0615481
68	4624	314432	8.2462113	4.0816551
69	4761	328509	8.3066239	4.1015659
70	4900	343000	8.3666003	4.1212853

n	n²	n³	√n	∛n
71	5041	357911	8.4261498	4.1408177
72	5184	373248	8.4852814	4.1601676
73	5329	389017	8.5440037	4.1793392
74	5476	405224	8.6023253	4.1983365
75	5625	421875	8.6602540	4.2171633
76	5776	438976	8.7177979	4.2358236
77	5929	456533	8.7749644	4.2543209
78	6084	474552	8.8317609	4.2726587
79	6241	493039	8.8881944	4.2908404
80	6400	512000	8.9442719	4.3088694
81	6561	531441	9.0000000	4.3267487
82	6724	551368	9.0553851	4.3444815
83	6889	571787	9.1104336	4.3620707
84	7056	592704	9.1651514	4.3795191
85	7225	614125	9.2195445	4.3968297
86	7396	636056	9.2736185	4.4140050
87	7569	658503	9.3273791	4.4310476
88	7744	681472	9.3808315	4.4479602
89	7921	704969	9.4339811	4.4647451
90	8100	729000	9.4868330	4.4814047
91	8281	753571	9.5393920	4.4979414
92	8464	778688	9.5916630	4.5143574
93	8649	804357	9.6436508	4.5306549
94	8836	830584	9.6953597	4.5468359
95	9025	857375	9.7467943	4.5629026
96	9216	884736	9.7979590	4.5788570
97	9409	912673	9.8488578	4.5947009
98	9604	941192	9.8994949	4.6104363
99	9801	970299	9.9498744	4.6260650
100	10000	1000000	10.0000000	4.6415888

SQUARE ROOT OF FRACTIONS

Fraction	Square Root
1/8	.3535
1/4	.5000
3/8	.6124
1/2	.7071
5/8	.7906
3/4	.8660
7/8	.9354

CONVERSION TABLES

MULTIPLY	BY	TO OBTAIN
Atmospheres (Std.) 760 MM of Mercury at 32°F.	14.696	Lbs./sq. inch
Atmospheres	76.0	Cms. of mercury
Atmospheres	29.92	In. of mercury
Atmospheres	33.90	Feet of water
Atmospheres	1.0333	Kgs./sq.cm.
Atmospheres	14.70	Lbs./sq. inch
Atmospheres	1.058	Tons/sq. ft.
Brit. Therm. Units	0.2520	Kilogram-calories
Brit. Therm. Units	777.5	Foot-lbs.
Brit. Therm. Units	0.000393	Horse-power-hrs.
Brit. Therm. Units	0.293	Watt-hrs.
BTU/min.	12.96	Foot-lbs./sec.
BTU/min.	0.02356	Horse-power
BTU/min.	0.01757	Kilowatts
BTU/min.	17.57	Watts
Calorie	0.003968	BTU
Centimeters	0.3937	Inches
Centimeters	0.03280	Feet
Centimeters	0.01	Meters
Centimeters	10	Millimeters
Centmtrs. of Merc.	0.01316	Atmospheres
Centimtrs. of merc.	0.4461	Feet of water
Centimtrs. of merc.	136.0	Kgs./sq. meter
Centimtrs. of merc.	27.85	Lbs./sq. ft.
Centimtrs. of merc.	0.1934	Lbs./sq. inch
Cubic feet	2.832×10^4	Cubic cms.
Cubic feet	1728	Cubic inches
Cubic feet	0.02832	Cubic meters
Cubic feet	0.03704	Cubic yards
Cubic feet	7.48052	Gallons U.S.
Cubic feet/minute	472.0	Cubic cms./sec.
Cubic feet/minute	0.1247	Gallons/sec.
Cubic foot water	62.4	Pounds @ 60°F.
Feet	30.48	Centimeters
Feet	12	Inches
Feet	0.3048	Meters
Feet	1/3	Yards
Feet of water	0.02950	Atmospheres
Feet of water	0.8826	Inches of mercury
Feet of water	0.03048	Kgs./sq. cm.

MULTIPLY	BY	TO OBTAIN
Feet of water	62.43	Lbs./sq. ft.
Feet of water	0.4335	Lbs./sq. inch
Feet/min.	0.5080	Centimeters/sec.
Feet/min.	0.01667	Feet/sec.
Feet/min.	0.01829	Kilometers/hr.
Feet/min.	0.3048	Meters/min.
Feet/min.	0.01136	Miles/hr.
Foot-pounds	0.001286	BTU
Gallons	3785	Cu. centimeters
Gallons	0.1337	Cubic feet
Gallons	231	Cubic inches
Gallons	128	Fluid ounces
Gallons	3.785	Liters
Gallons water	8.35	Lbs.water @60°F.
Horse-power	42.44	BTU/min.
Horse-power	33,000	Foot-lbs./min.
Horse-power	550	Foot-lbs./sec.
Horse-power	0.7457	Kilowatts
Horse-power	745.7	Watts
Horse-power (boiler)	33,479	BTU/hr.
Horse-power (boiler)	9.803	Kilowatts
Horse-power-hours	2547	BTU
Horse-power-hours	0.7457	Kilowatt-hours
Inches	2,540	Centimeters
Inches	25.4	Millimeters
Inches	0.0254	Meters
Inches	0.0833	Foot
Inches of mercury	0.03342	Atmospheres
Inches of mercury	1.133	Feet of water
Inches of mercury	13.57	Inches of water
Inches of mercury	70.73	Lbs./sq. ft.
Inches of mercury	0.4912	Lbs./sq. inch
Inches of water	0.002458	Atmospheres
Inches of water	0.07355	In. of mercury
Inches of water	0.5781	Ounces/sq. inch
Inches of water	5.202	Lbs./sq. foot
Inches of water	0.03613	Lbs./sq. inch
Kilowatts	56.92	BTU/min.
Kilowatts	1.341	Horse-power
Kilowatts	1000	Watts
Kilowatt-hours	3415	BTU

MULTIPLY	BY	TO OBTAIN
Liters	0.2642	Gallons
Liters	2.113	Pints (liq.)
Liters	1.057	Quarts (liq.)
Meters	100	Centimeters
Meters	3.281	Feet
Meters	39.37	Inches
Meters	1000	Millimeters
Meters	1.094	Yards
Ounces (fluid)	1.805	Cubic inches
Ounces (fluid)	0.02957	Liters
Ounces/sq. inch	0.0625	Lbs./sq. inch
Ounces/sq. inch	1.73	Inches of water
Pints	0.4732	Liter
Pounds (avoir.)	16	Ounces
Pounds of water	0.01602	Cubic feet
Pounds of water	27.68	Cubic inches
Pounds of water	0.1198	Gallons
Pounds/sq. foot	0.01602	Feet of water
Pounds/sq. foot	0.006945	Pounds/sq. inch
Pounds/sq. inch	0.06804	Atmospheres
Pounds/sq. inch	2.307	Feet of water
Pounds/sq. inch	2.036	In. of mercury
Pounds/sq. inch	27.68	Inches of water
Temp.(°C.) + 273	1	Abs. temp. (°C.)
Temp.(°C.) + 17.78	1.8	Temp. (°F.)
Temp.(°F.) + 460	1	Abs. temp. (°F.)
Temp.(°F.) − 32	5/9	Temp. (°C.)
Therm	100,000	BTU
Tons(long)	2240	Pounds
Ton, Refrigeration	12,000	BTU/hr.
Tons (short)	2000	Pounds
Watts	3.415	BTU
Watts	0.05692	BTU/min.
Watts	44.26	Foot-pounds/min.
Watts	0.7376	Foot-pounds/sec.
Watts	0.001341	Horse-power
Watts	0.001	Kilowatts
Watt-hours	3.415	BTU/hr.
Watt-hours	2655	Foot-pounds
Watt-hours	0.001341	Horse-power hrs.
Watt-hours	0.001	Kilowatt-hours

TEMPERATURE CONVERSION TABLE

Use this table to convert listed temperatures from either Fahrenheit to Celsius, or Celsius to Fahreneit. For temperatures not listed, use the following formulas;

$$\text{Fahrenheit} = (C \times 1.8) + 32 = F \qquad \text{Celsius} = (F-32) \times .555 = C$$

C	F	C	F	C	F	C	F	C	F	C	F
-30	-22.0	20	68.0	70	158.0	120	248.0	170	338.0		
-29	-20.2	21	69.8	71	159.8	121	249.8	171	339.8		
-28	-18.4	22	71.6	72	161.6	122	251.6	172	341.6		
-27	-16.6	23	73.4	73	163.4	123	253.4	173	343.4		
-26	-14.8	24	75.2	74	165.2	124	255.2	174	345.2		
-25	-13.0	25	77.0	75	167.0	125	257.0	175	347.0		
-24	-11.2	26	78.8	76	168.8	126	258.8	176	348.8		
-23	-9.4	27	80.6	77	170.6	127	260.6	177	350.6		
-22	-7.6	28	82.4	78	172.4	128	262.4	178	352.4		
-21	-5.8	29	84.2	79	174.2	129	264.2	179	354.2		
-20	-4.0	30	86.0	80	176.0	130	266.0	180	356.0		
-19	-2.2	31	87.8	81	177.8	131	267.8	181	357.8		
-18	-0.4	32	89.6	82	179.6	132	269.6	182	359.6		
-17	+1.4	33	91.4	83	181.4	133	271.4	183	361.4		
-16	3.2	34	93.2	84	183.2	134	273.2	184	363.2		
-15	5.0	35	95.0	85	185.0	135	275.0	185	365.0		
-14	6.8	36	96.8	86	186.8	136	276.8	186	366.8		
-13	8.6	37	98.6	87	188.6	137	278.6	187	368.6		
-12	10.4	38	100.4	88	190.4	138	280.4	188	370.4		
-11	12.2	39	102.2	89	192.2	139	282.2	189	372.2		
-10	14.0	40	104.0	90	194.0	140	284.0	190	374.0		
-9	15.8	41	105.8	91	195.8	141	285.8	191	375.8		
-8	17.6	42	107.6	92	197.6	142	287.6	192	377.6		
-7	19.4	43	109.4	93	199.4	143	289.4	193	379.4		
-6	21.2	44	111.2	94	201.2	144	291.2	194	381.2		
-5	23.0	45	113.0	95	203.0	145	293.0	195	383.0		
-4	24.8	46	114.8	96	204.8	146	294.8	196	384.8		
-3	26.6	47	116.6	97	206.6	147	296.6	197	386.6		
-2	28.4	48	118.4	98	208.4	148	298.4	198	388.4		
-1	30.2	49	120.2	99	210.2	149	300.2	199	390.2		
0	32.0	50	122.0	100	212.0	150	302.0	200	392.0		
1	33.8	51	123.8	101	213.8	151	303.8	201	393.8		
2	35.6	52	125.6	102	215.6	152	305.6	202	395.6		
3	37.4	53	127.4	103	217.4	153	307.4	203	397.4		
4	39.2	54	129.2	104	219.2	154	309.2	204	399.2		
5	41.0	55	131.0	105	221.0	155	311.0	205	401.0		
6	42.8	56	132.8	106	222.8	156	312.8	206	402.8		
7	44.6	57	134.6	107	224.6	157	314.6	207	404.6		
8	46.4	58	136.4	108	226.4	158	316.4	208	406.4		
9	48.2	59	138.2	109	228.2	159	318.2	209	408.2		
10	50.0	60	140.0	110	230.0	160	320.0	210	410.0		
11	51.8	61	141.8	111	231.8	161	321.8	211	411.8		
12	53.6	62	143.6	112	233.6	162	323.6	212	413.6		
13	55.4	63	145.4	113	235.4	163	325.4	213	415.4		
14	57.2	64	147.2	114	237.2	164	327.2	214	417.2		
15	59.0	65	149.0	115	239.0	165	329.0	215	419.0		
16	60.8	66	150.8	116	240.8	166	330.8	216	420.8		
17	62.6	67	152.6	117	242.6	167	332.6	217	422.6		
18	64.4	68	154.4	118	244 4	168	334.4	218	424.4		
19	66.2	69	156.2	119	246.2	169	336.2	219	426.2		

EARTHWORK

A job site must first be cleared and grubbed by removing grass, rock, trees, etc., and then defined as to existing elevations from a bench mark reference. The existing cleared elevation must then be raised or lowered by bringing in fill, or removing earth.

Excavating is measured in cubic yards, wherein one cubic yard contains 27 cubic feet (3'-0" x 3'-0" x 3'-0").

The conditions of land varies and different soil types must be handled properly to prevent problems. To prevent cave-ins, an "angle of repose" must be applied for safe working conditions.

ANGLE OF REPOSE
FOR SLOPING OF SIDES OF EXCAVATIONS

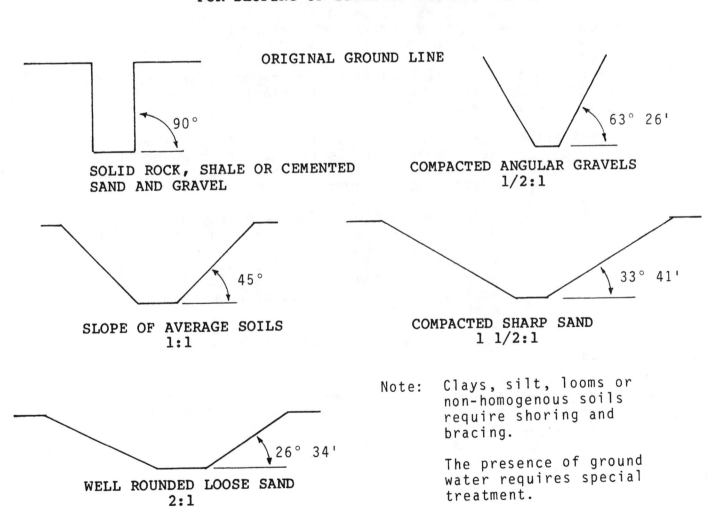

ORIGINAL GROUND LINE

90°

SOLID ROCK, SHALE OR CEMENTED SAND AND GRAVEL

63° 26'

COMPACTED ANGULAR GRAVELS
1/2:1

45°

SLOPE OF AVERAGE SOILS
1:1

33° 41'

COMPACTED SHARP SAND
1 1/2:1

26° 34'

WELL ROUNDED LOOSE SAND
2:1

Note: Clays, silt, looms or non-homogenous soils require shoring and bracing.

The presence of ground water requires special treatment.

FIG. 1

In other cases shoring and bracing must be used. Refer to WALKER'S ESTIMATOR'S REFERENCE BOOK, pages 195-200. When ground water conditions prevail a method of de-watering must be applied, such as a Well Point System. See WALKER'S EST. REFERENCE BOOK, pages 7.87-7.88.

When soil conditions are unstable and do not provide adequate bearing for normal footings, piling must be used. The common types of piling are:

1. Reinforced concrete; precast & driven.
2. Reinforced concrete; precast, prestressed.
3. Concrete cast in driven steel casing.
4. Steel H Beam.
5. Timbers.

Pilings are driven into the ground vertically until they reach a solid foundation. The load is determined by the number of blows required by the pile hammer to drive the pile the last few inches, or until the pile is driven to "refusal". See WALKER'S ESTIMATING REFERENCE BOOK, pages 7.64-7.75, and PLACING REINFORCING BARS, page 3-2.

When excavating and using an "angle of repose", (See WALKER'S EST. REF. BOOK, page 7.88, for illustration), the method used to estimate the amount of earth to be removed is averaging.

EXAMPLE: <u>Given</u>: A 100 ft x 50 ft x 6 ft deep basement is to be dug at a site that is composed of compacted sharp sand. What is the amount of earth to be removed?

SOLUTION: Referring to Fig. 1 "Angle of Repose", the diagram shows compacted sharp sand to have an angle of repose of 1-1/2' : 1'. Meaning that for each foot down, it is required to go out 1-1/2 feet. The first number is always the distance out and the second number the distance down.

Drawing the two elevations of the proposed excavation as follows:

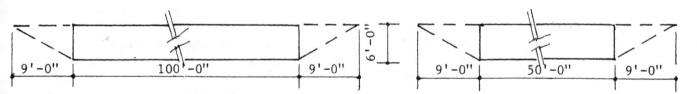

FIG. 2

The two triangles on either side of the rectangle are symmetrical (equal size).

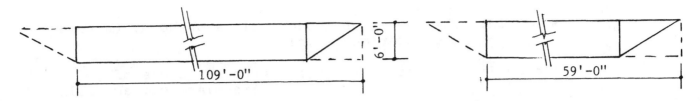

FIG 3

Therefore, a left-side triangle can be moved to the right side thus forming a rectangle volume 109.0' x 59.0' x 6.0' deep equaling 38,586 cubic feet. This method of averaging leaves out one corner which is a pyramid. The formula for this corner is, $\frac{B \times B \times H}{3} = \frac{9' \times 9' \times 6'}{3} =$
162 cu ft + 38,586 cu ft. 38,748 cu ft / 27 = 1435.11 cu yd. See Figure 3.

If this problem was shored on one end on the short side, the left-side triangle may be moved to the right side of the right drawing (Figure 2) and the rectangle volume would equal 100' x 59' x 6' = 35,400 cu ft. There are no triangles in the left drawing because the sides are shored.

When earth is removed, the digging causes the volume of the earth to be greater than the original compacted soil; this is known as expansion. If, in the above example, it was stated that the earth expanded 20%, then the total volume must be mulitplied by 1.20 giving a final volume of 1722.13 cu yds (1435.11 x 1.20 = 1722,13). Sometimes the test writers will give a condition of expansion, at other times they will not. Remember that when expansion is asked for, the total volume must be muultiplied by the expansion factor plus 100%.

Earth may also be compacted. When compacting earth, the final result is the volume remaining after compaction. When the earth is compacted for a 25% loss in volume, 75% remains. If 600 cu yds of earth are brought to the job site and compacted to 75%, the remaining volume would be;

600 cu yd x .75 = 450 cu yd.

Here's how the test writers would word the problem;

Given: The total volume of in-place compacted fill is 450 cu yd, and the loose fill causes a 25% loss in volume. The total volume of loose fill required is _____ .

SOLUTION: The volume of loose fill is unknown; therefore the equation reads:

(loose fill) x (remaining fill) = total volume of in-place compact fill

Where: Remaining fill = (100% - % loss in volume)

therefore; x x (100% - 25%) = 450 cu yds
x x .75   = 450 cu yds
x = 450/75   = 600 cu yds

Earth is compacted with mechanical rollers, either sheepsfoot rollers, rubber tire rollers, or drum rollers. WALKER'S ESTIMATING REFERENCE BOOK, page 7.53, refers to the use of rolling. Know how to use this table as questions have been asked from it.

Estimating the Cost of Excavating

Test writers are beginning to ask more questions relating to cost estimating. Like other specific tasks, excavating has come under closer scrutiny. Estimating the cost of excavating requires an accurate knowledge of the type of soil in question. Soils may be classified into four general categories for practical purposes:

Sandy or light. Easily shoveled, requiring no loosening.
Medium light or ordinary. Earth easily loosened by pick or jabbing shovel. When power equipment is considered, no preliminary loosening is required.
Heavy, hard or clay. Requires heavy pick work to loosen, or may be dug with heavy power shovel without preliminary loosening.
Hardpan. Requires light blasting and heavy excavating machinery. Soils not classified here, such as rock, must be carefully considered and should be handled under separate agreement. Typically, Chart 1 gives a production scale according to depth for hand excavating pipe trenching.

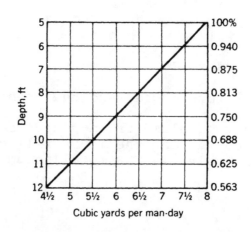

CHART 1
PRODUCTION CHART FOR HAND EXCAVATING

The earth's terrain is not level; therefore, reference points are established by land surveyors known as identification marks or bench marks, which are placed on permanent concrete markers, at locations at cross roads, or other permanently fixed locations. A bench mark serves as a datum (reference position) on a building site. From the bench mark, other elevations may be established at a job site to determine heights at various points for relocating earth and determining the amount of fill required or earth to be removed. If four points of elevation were known at a site the average elevation can be determined:

EXAMPLE:

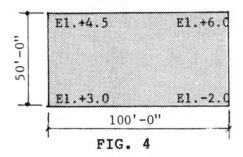

FIG. 4

By using averaging, add the four points [(+4.5) + (+6.0) + (+3.0) + (-2.0)] = 11.5 divided by 4 (averaging) = +2.875 which would be the average elevation if the ground were uniformly graded. If a new elevation were to be established the volume of fill or removal of earth can be determined using the **FORMULA FOR VOLUME** as follows:

$$V = L \times W \times (E_f - E_a)$$

Where:

V = Volume
L = Length
W = Width
E_f = Elevation Final
E_a = Elevation Average

Using the example in Fig. 4, assume a new final elevation of +1.5, and apply the above formula;

$$V = 100 \times 50 \times [+1.5 - (+2.875)]$$

Then, applying the rule of signs and working the parenthesis first;

$$V = 100 \times 50 \times [+1.5 - 2.875]$$
$$V = 100 \times 50 \times (-1.375)$$
$$V = -6875 \text{ cu ft} / 27$$
$$V = -254.6 \text{ cu yds}$$

(the minus sign indicates earth to be removed)

Should the new final elevation be +3.5 the formula reads:

$$V = 100 \times 50 \times [+3.5 - (+2.85)]$$
$$V = 100 \times 50 \times (+.65)$$
$$V = 3250 \text{ cu ft}/27 = 120.37 \text{ cu yds}$$

(plus indicates fill to be brought in)

Another method of determining elevations is by drawing a vertical scale with a zero axis. See Fig. 5 below.

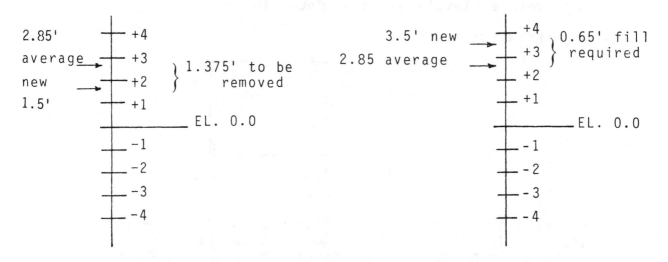

FIG. 5

The existing elevation and new elevation may be located on the scale and it can be determined whether the final result is fill or earth removal.

A parcel of land may not be uniformly level and an average elevation can be determined in order to know the amount of fill or earth removal in order to establish a new elevation. The method for determining the existing elevations is known as topography (grafting the top surface). The procedure is to locate various points on the site, then find their elevations.

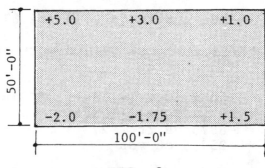

FIG. 6

Adding all six points [(+5) + (+3) + (+1) + (+1.5) + (−1.75) + (−2) = +6.75] to determine the average elevation. Dividing by the six points (6.75 / 6 = 1.125) the average elevation is +1.125. A topographic survey is shown on p 7.15 of WALKER'S EST. REF. BOOK using the cross sectional method of survey. The test writers have used variations of this type of survey in previous examinations. Any particular grid or series of grids may be taken from the illustration on page 7.15 of WALKER'S EST. REFERENCE BOOK and existing and new elevations can be determined. A typical example is as follows:

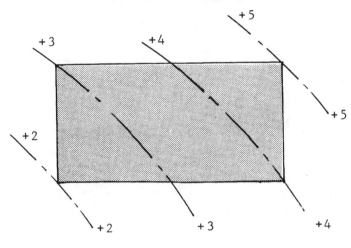

FIG. 7

Six points are established on the site in Fig. 7. Averaging (+2 +3 +4 +5 +4 +3 = 21), divided by 6 = 3.5 (average elevation).

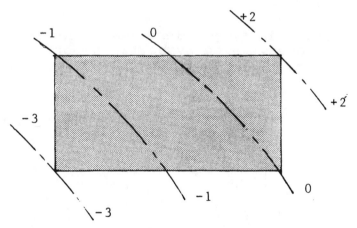

FIG. 8

Six points are established on the site in Fig. 8. Averaging [+ (−3) + (−1) + (0) + (+2) + (0) + (−1) = −3], divided by 6 = −0.5. (average elevation).

In both Fig. 7 and Fig. 8, a new elevation of +2.0 is to be established.

Applying the formula: $V = L \times W \times (E_f - E_a)$.

Fig. 7 $V = 100 \times 50 \times [(+2) - (+3.5)]$
$V = 100 \times 50 \times (-1.5) = -7500$
$-7500/27 = -278$ cu yds.
(minus indicates earth to be removed)

Fig. 8 $V = 100 \times 50 \times (+2) - (-.5) =$
$100 \times 50 \times (+2.5) = +12,500$
$+ 12,500/27 = +463$ cu yds.
(plus indicates fill required)

Graphically:

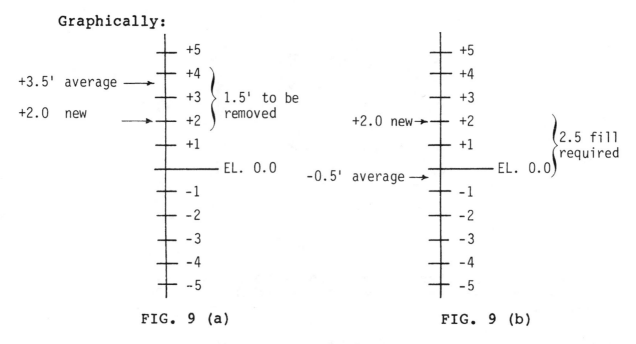

FIG. 9 (a)   FIG. 9 (b)

In previous examinations, problems have been given with uniform graded incline slopes, which symbolize a portion of a topography as illustrated on page 7.15 of WALKER'S ESTIMATOR'S REFERENCE BOOK.

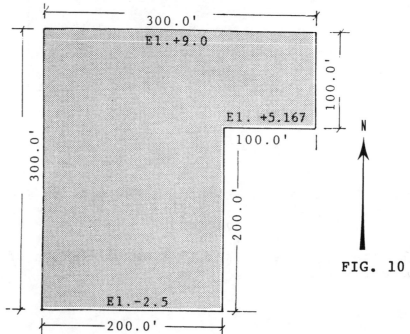

FIG. 10

Visualizing Fig. 10 in isometric.

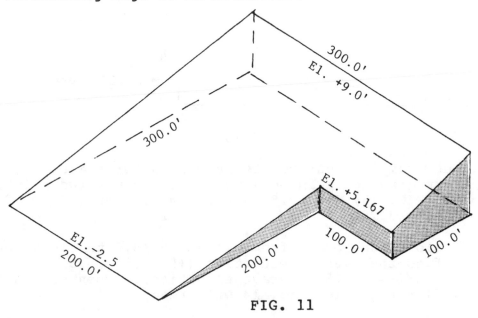

FIG. 11

The problem would read as follows:

EXAMPLE: <u>Given</u>: The site slopes uniformly from elevation +9 along the northernmost side to -2.5 along the south side.

It is desired to grade the site to a uniform flat elevation of +1.50'. The total amount of earth to be removed is _____ .

Step 1. Each rectangular area must be calculated separately.

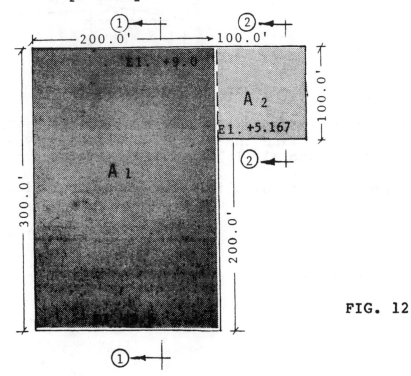

FIG. 12

Step 2. Visualizing Al from an elevation view 1-1.

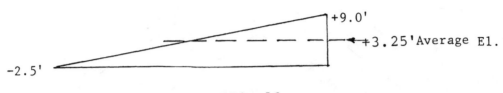

FIG. 13

In previous examples four points were used for averaging.
With uniform grading it is only necessary to use two points.
Refering to FIG. 13; (+9 -2.5) = 6.5/2 = 3.25 or (+9 +9 -2.5
-2.5) = 18/4 = 3.25.

The grade slopes uniformly from elevation +9.0 to elevation
-2.5. To find the average elevation (+9 -2.5 = 6.5)
averaging 6.5/2 = +3.25 (average elevation). When graded to
a uniform flat elevation Fig. 14 will read;

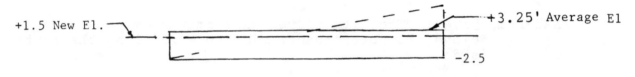

FIG. 14

Visualizing A_2 from elevation view 2-2.

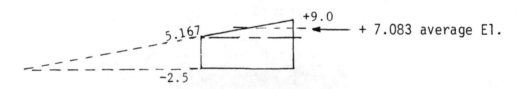

FIG. 15

The grade slopes uniformly from elevation +9.0 to elevation
+5.167. To find the average elevation (+9 +5.167 = +14.167)
and averaging +14.167/2 = +7.083 (average elevation).

When graded to a uniform flat elevation Fig. 16 will read,

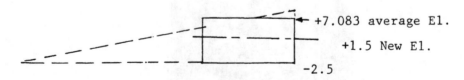

FIG. 16.

$A_1;$ $V = 300 \times 200 \times [+1.5 - (+3.25)]$
$300 \times 200 \times (-1.75) = -105,000$ cu ft
$-105,000/27 = -3888.9$ cu yds
(minus indicates earth to be removed)

$A_2;$ $V = 100 \times 100 \times [+1.5 - (+7.083)]$
$100 \times 100 \times (-5.583) = -55,830$ cu ft
$-55,830/27 = -2068$ cu yds
(minus indicates earth to be removed)

$-3888.9 + (-2068) = -5956.9$ cu yds of earth to be removed.

There are times when the test writers will leave out elevations on the uniformly graded incline slope problems.

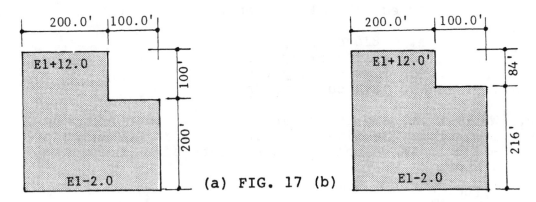

(a) FIG. 17 (b)

In Fig. 17(a) the unknown elevation is at an equal third. The total difference in elevation is from +12 to -2, a total of 14 ft, 14/3 = 4.667' down from the top; therefore, +12 - 4.667' = +7.333 missing elevation.

In Fig. 17(b) the missing elevation is not at an equal third; therefore, the slope of the incline must be calculated. Slope is the ratio of rise to run. Rise is 14 ft and run is 300 ft, 14/300 = .047' of rise for each ft of run. For 84' of run, multiply 84 x .047 which equals 3.95 ft of rise. Subtracting; 12 - 3.95 = +8.05 missing elevation.

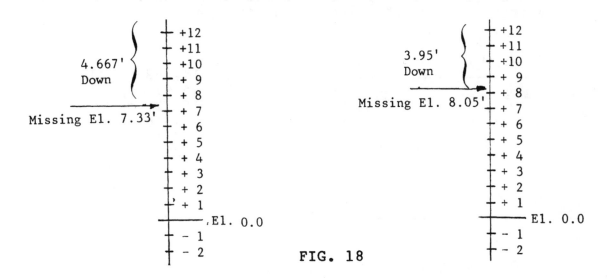

FIG. 18

EXCAVATION PROBLEM NO. 1

1. GIVEN: A 200' long, 10' deep trench is to be excavated with the sides sloped from the base according to minimum OSHA Regulations. No shoring is to be used on the 200' long sides. The soil is "average" and the base width of the trench is 6'. The ends of the trench are to be shored and will not be sloped.

   According to the procedures for estimating quantities of excavation discussed in WALKER'S ESTIMATOR'S REFERENCE BOOK; the total volume of excavation of the trench is_____

   Select the closest answer.

   A.   815 cubic yards
   B. 1185 cubic yards
   C. 1244 cubic yards
   D. 1926 cubic yards

2. What is the amount of earth to be removed from the project in the diagram, with a uniform elevation of -3.75'. The slope of the existing site is uniformly graded.

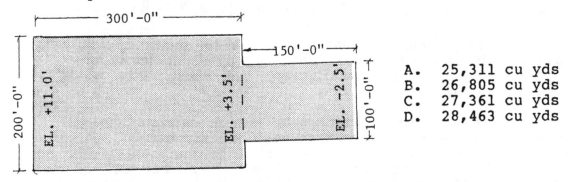

   A.   25,311 cu yds
   B.   26,805 cu yds
   C.   27,361 cu yds
   D.   28,463 cu yds

3. What is the amount of earth to be removed from the site in the diagram if a uniform elevation of +18.00' is desired. Select the closest answer.

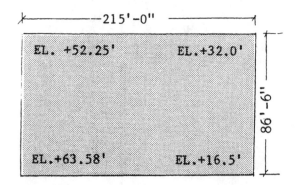

   A.   14,862 cu yds
   B.   15,911 cu yds
   C.   16,325 cu yds
   D.   17,540 cu yds

EXCAVATION PROBLEM NO. 2

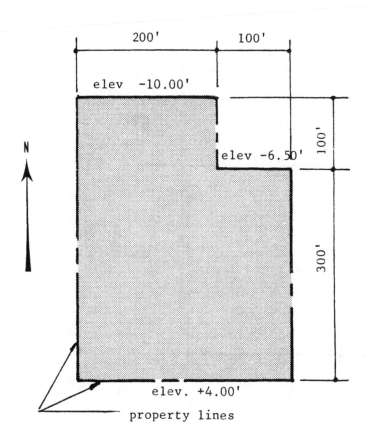

GIVEN: The lot shown above slopes uniformly from elevation
-10.00' along the northernmost line to +4.00' along
the south line. The lot is to be filled and graded
to a uniform flat elevation of 0.00'. No earth is to
be removed from the property lines. No allowance is
to be made for compaction of fill brought to the lot.

The total volume of fill required from outside of the
property lines is_____.

    A. less than 9,000 cubic yards
    B. between 9,000 and 10,000 cubic yards
    C. between 10,000 and 11,000 cubic yards
    D. between 11,000 and 12,000 cubic yards

EXCAVATION PROBLEM NO. 3

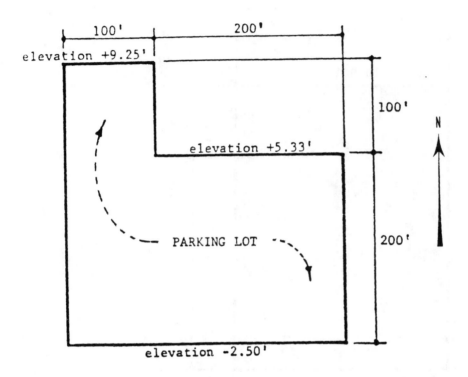

GIVEN: The parking lot shown above slopes uniformly from elevation +9.25' along the northernmost side to -2.50' along the south side.

It is desired to grade the parking lot to a uniform flat elevation of +1.50'. The total net volume of earth to be removed from the lot is _____ . No fill from outside the lot is to be used. Select the closest answer.

    A. 1956 cubic yards
    B. 2682 cubic yards
    C. 5100 cubic yards
    D. 7511 cubic yards

EXCAVATION PROBLEM NO. 4

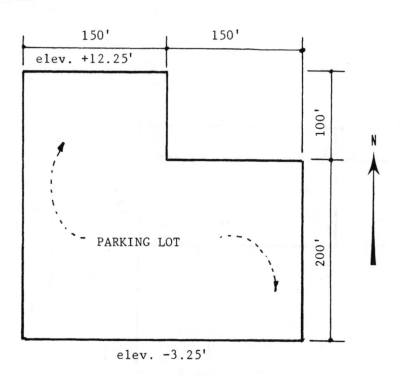

elev. +12.25'

150'   150'

100'

200'

N

PARKING LOT

elev. -3.25'

GIVEN: The parking lot shown above slopes uniformly from elevation +12.25' along the northmost side to -3.25' along the south side.

It is desired to grade the parking lot to a uniform flat elevation of +2.25'. The total net volume of earth to be removed from the lot is _____. No fill from outside the lot is to be used. Select the closest answer.

        A. 2934 cubic yards
        B. 3390 cubic yards
        C. 5100 cubic yards
        D. 6845 cubic yards

EXCAVATION PROBLEM NO. 5

1. From WALKER'S ESTIMATOR'S, p 7.15, what is the amount of earth that will be removed in cubic yards from the area, E-1 to J-1, J-1 to J-2, J-2 to H-2, H-2 to H-3, H-3 to F-3, F-3 to F-4, F-4 to G-4, G-4 to G-5, G-5 to E-5, E-5 to E-1?

    This could be the type of problem you would receive on the exam.

2.

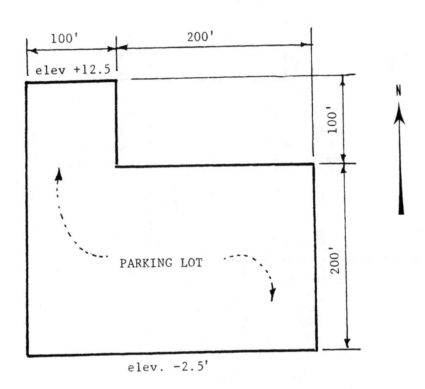

elev. -2.5'

GIVEN: The parking lot shown above slopes uniformly from elevation +12.5' along the northernmost side to -2.5' along the south side.

It is desired to grade the parking lot to a uniform flat elevation of +1.5'. The total volume of earth to be removed from the lot is _____ . No fill from outside the lot is to be used. Select the closest answer.

    A. 1956 cubic yards
    B. 2750 cubic yards
    C. 5370 cubic yards
    D. 7511 cubic yards

EXCAVATION PROBLEM NO. 6

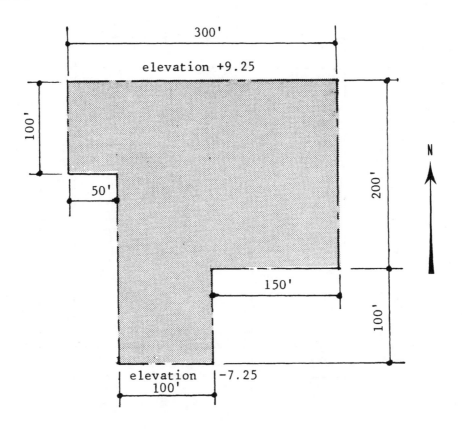

GIVEN: The lot shown above slopes uniformly from elevation
        +9.25' along the northernmost line to -7.25' along
        the south line. The lot is to be filled and graded
        to a uniform flat elevation of +3.5'. No earth is to
        be removed from the property lines. No allowance is
        to be made for compaction of fill brought to the lot.

The total volume of fill required from outside of the
property lines is _____.

    A. less than 2000 cubic yards
    B. between 2000 and 3000 cubic yards
    C. between 3000 and 4000 cubic yards
    D. between 4000 and 5000 cubic yards

EXCAVATION PROBLEM NO. 7

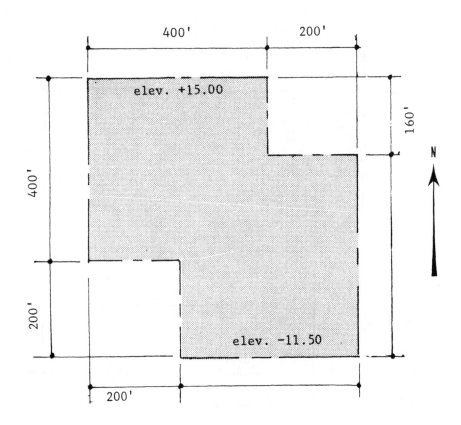

GIVEN: The lot shown above slopes uniformly from the
         elevation of +15.00' along the northernmost line
         to -11.50' along the southernmost line.

It is desired to grade the parking lot to a uniform flat
elevation of -1.5'. The total net volume of earth to be removed
from the lot is _____. No fill from outside the lot is to be
used. Select the closest answer.

        A.   29,996 cubic yards
        B.   32,768 cubic yards
        C.   34,024 cubic yards
        D.   36,195 cubic yards

EXCAVATION PROBLEM NO. 8

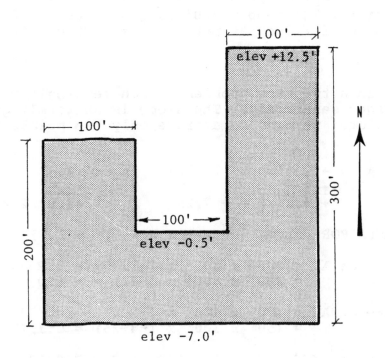

GIVEN: The lot shown above slopes uniformly from elevation +12.5' along the northernmost line to -7.0'along the southernmost line.

It is desired to grade the lot to a uniform flat elevation of +2.5'. The total net volume of fill required to be delivered to the site is _____. Select the closest answer.

    A. 2962 cubic yards
    B. 3888 cubic yards
    C. 4260 cubic yards
    D. 4999 cubic yards

ANSWER SHEET

EXCAVATION QUIZ NO. 1

1. From OSHA p 132, average soil has a 1:1 angle of repose. Therefore the trench is 10' deep and the top of the excavation is 26' wide and the bottom is 6' wide.

Using averaging, 26' + 6' = 32'/2 = 16' average width. 16' wide x 10' deep x 200' long = 32,000 cu ft. 32,000 cu ft/27 cu ft per cu yd = 1185 cu yds.

ANSWER B

2. This is a two step problem. Each rectangle must be averaged separately. The slope is uniformly graded; therefore, we need to average only two points in each step.

$$A_1 = \begin{array}{r} +11.0' \\ +\ 3.5' \\ \hline +14.5'/2 = +7.25' \end{array} \qquad A_2 = \begin{array}{r} +3.5' \\ -2.5' \\ \hline +1.0'/2 = +.5' \end{array}$$

FORMULA FOR VOLUME: $V = L \times W \times (E - E)$

Volume of A_1 = 200' x 300' x [-3.75-(+ 7.25)]
= 200' x 300' x (-11) = - 660,000 cu ft.

Volume of A_2 = 150' x 100' x [-3.75 - (+.5)]
= 150' x 100' x (-4.25) = -63,750 cu ft.

$A_1 + A_2$ = -660,000 + (-63,750) = 723,750/27 = 26,805.556 cu yds of dirt to be removed.

ANSWER B

3. The elevation is not uniformly graded; therefore, four points must be averaged.

$$\begin{array}{r} +52.25 \\ +32.00 \\ +63.58 \\ +16.50 \\ \hline +164.33/4 = 41.1 \text{ (average elevation)} \end{array}$$

FORMULA FOR VOLUME: $V = L \times W \times (E_f - E_a)$

V = 215' x 86.5' x [+18 - (+41.1)] = 215' x 86.5' x (-23.1) = -429,602.25/27 = -15,911.2 cu yds to be removed.

ANSWER B

ANSWER SHEET

EXCAVATION PROBLEM NO. 2

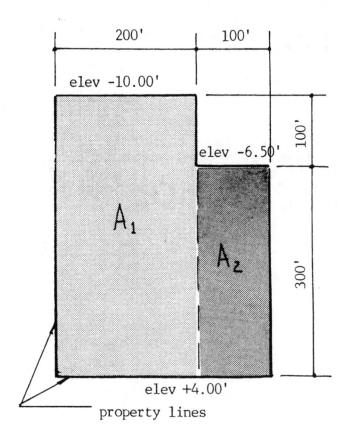

FORMULA FOR VOLUME: $V = L \times W \times (E_f - E_a)$

L = length
W = width
E_f = elevation final
E_a = elevation average

1. Areas must be divided into equal rectangles $A_1 - A_2$

2. Average elevation must be found:

A = -10   A = -6.5
    + 4   +4.0
    ―――   ―――
    -6/2 = -3   -2.5/2 = -1.25

3. Volume of A_1 = 400 x 200 x [0-(-3)]
   = 400 X 200 X (+3)
   = +240,000 cu ft/27 cu ft per cu yd
   = +8888.8 cu yds.

Note: Do not apply sign until multiplication process is completed.

4. Volume of A_2 = 300 x 100 x [0-(-1.250)]
   = 300 X 100 X (+1.250)
   = +37,500 cu ft/27 cu ft per cu yd
   = +1389 cu yds.

5. Both signs are +; therefore, add,   + 8888.8
                                         + 1388.0
                                         +10,277.0 cu yds

Note: If the final answer is + then fill is to be added, if the final answer is - then earth is to be removed. Be careful of the signs. If A_1 is plus and A_2 is minus fill is taken from one area and added to the other, the answer being the difference.

ANSWER C

ANSWER SHEET

EXCAVATION PROBLEM NO. 3

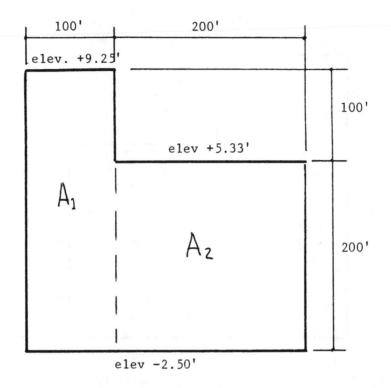

FORMULA FOR VOLUME: $V = L \times W \times (E_f - E_a)$

1. Find average elevation:

$$A_1 = \begin{array}{r} +9.25 \\ \underline{-2.50} \\ +6.75/2 = +3.375 \end{array} \qquad A_2 = \begin{array}{r} +5.33 \\ \underline{-2.50} \\ +2.83/2 = +1.415 \end{array}$$

2. Volume of A_1 = 300' x 100' x [+1.5 - (+3.375)] =
   = 300' x 100' x (-1.875)
   = -56,250 cu ft/27 = -2083.3 cu yds

3. Volume of A_2 = 200' x 200' x [+1.5 - (+1.415)]
   = 200' x 200' x (+.085)
   = +3400 cu ft/27 = +126 cu yds

4. Signs being opposite, subtract:
   $$\begin{array}{r} -2083.3 \text{ cu yds} \\ \underline{+\ 126.0} \text{ cu yds} \\ -1957.3 \text{ cu yds of earth} \\ \text{to be removed} \end{array}$$

ANSWER A

ANSWER SHEET

EXCAVATION PROBLEM NO. 4

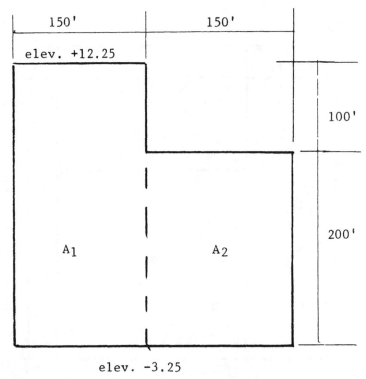

FORMULA FOR VOLUME: $V = L \times W \times (E_f - E_a)$

1. Solve for missing elevation: From +12.25 to -3.25 is a
    total of 15.5 ft. 15.5'/3 equal parts = 5.167';
    therefore, +12.25 - 5.167 = EL +7.083.

2. Find average elevation:

$A_1 = \begin{array}{r} +12.25 \\ - 3.25 \\ \hline + 9.0/2 \end{array} = +4.5$ $A_2 = \begin{array}{r} +7.085 \\ -3.250 \\ \hline +3.833/2 \end{array} = +1.917$

3. Volume of A_1 = 150' x 300' x [+2.25 - (+4.5)]
                  = 150' x 300' x (-2.25)
                  = -101,250 cu ft / 27 = -3750 cu yds

4. Volume of A_2 = 150' x 200' x [+2.25 - (+1.917)]
                  = 150' x 200' x (+.333)
                  = +9990 cu ft / 27 = +370 cu yds

5. Signs being opposite, subtract: $\begin{array}{r} -3750 \text{ cu yds} \\ + 370 \text{ cu yds} \\ \hline -3380 \text{ cu yds to be} \\ \text{removed} \end{array}$

                                      ANSWER B

ANSWER SHEET

EXCAVATION PROBLEM NO. 5

1. Homework Problem from <u>Walker's Estimating</u>, page 7.15.

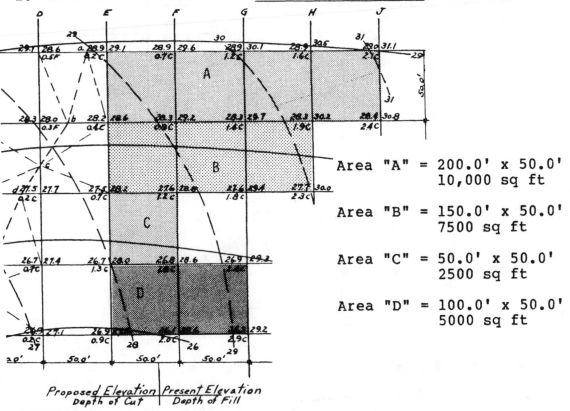

Area "A" = 200.0' x 50.0'
            10,000 sq ft

Area "B" = 150.0' x 50.0'
            7500 sq ft

Area "C" = 50.0' x 50.0'
            2500 sq ft

Area "D" = 100.0' x 50.0'
            5000 sq ft

AVERAGING:

"A"	"B"	"C"	"D"
.2	.4	.7	1.3
.7	.9	1.2	1.8
1.2	1.4	1.3	2.4
1.6	1.9	1.8	2.9
2.1	2.3	5.0/4	2.0
2.4	1.8	= 1.25	.9
1.9	1.2		11.3/6
1.4	.7		= 1.883
.9	10.6/8		
.4	= 1.325		
12.8/10			
= 1.28			

```
"A"  =  10,000 sq ft x 1.28  ft = 12,800.0 cu ft
"B"  =   7,500 sq ft x 1.325 ft =  9,937.5 cu ft
"C"  =   2,500 sq ft x 1.25  ft =  3,125.0 cu ft
"D"  =   5,000 sq ft x 1.883 ft =  9,415.0 cu ft
```

35,277.5 cu ft/27 = 1306.574 cu yds of earth to be removed.

ANSWER SHEET

EXCAVATION PROBLEM NO. 5-2

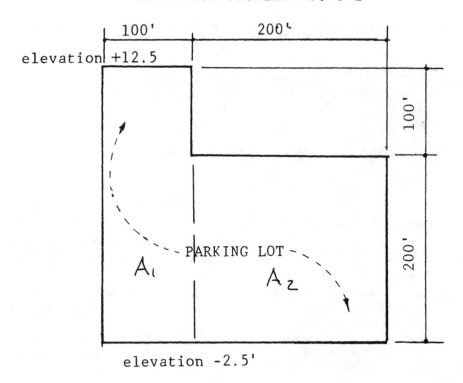

FORMULA FOR VOLUME: $V = L \times W \times (E_f - E_a)$

1. Solve for missing elevation: From +12.5 to -2.5 is a total of 15 ft. Rise per foot = 15/300 = .05 ft of rise per ft of run; .05 x 200' = 10 ft of rise from -2.5 = EL +7.5, or missing EL is at the 1/3 point, therefore; 15 ft rise/3 = 5 ft difference, +12.5' - 5' = EL +7.5'.

2. Find average elevation:

A_1 = +12.5' A_2 = +7.5'
       - 2.5'   -2.5'
      +10.0'/2 = +5'   +5.0'/2 = +2.5'

3. Volume of A_1 =   300' x 100' x [+1.5 - (+5)]
               =   300' x 100' x (-3.5)
               = -105,000 cu ft / 27 = -3888.9 cu yds

4. Volume of A_2 =   200' x 200' x [+1.5 - (+2.5)]
               =   200' x 200' x (-1)
               = -40,000 cu ft / 27 = -1481.48 cu yds.

6. Both signs being minus, we add: -3888.9 cu yds
                                      -1481.8 cu yds
                                      -5370.7 cu yds of earth
                                            to be removed.

ANSWER C

ANSWER SHEET

EXCAVATION QUIZ NO. 6

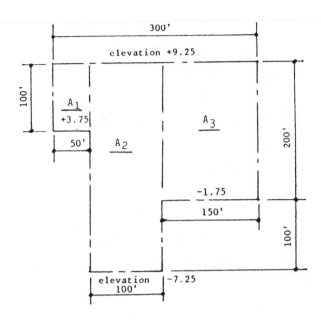

FORMULA FOR VOLUME: $V = L \times W \times (E_f - E_a)$

1. Solve for missing elevation: From +9.25 to -7.25 there
    is a total of 16.5'. The property is divided in three
    equal 100' lengths; therefore, 16.5/3 = 5.5' per 100'.
    A_1 = 9.25 - 5.5 = EL +3.75 A_3 = 9.25 - 11 = EL -1.75

2. Find average elevation:

    A_1 = +9.25 A_2 = +9.25 A_3 = +9.25
         +3.75   -7.25   -1.75
         +13.00/2 = +6.5   +2.0/2 = +1.0   +7.5/2 = +3.75

3. Volume of A_1 = 100' x 50' x [+3.5 - (+6.5)]
                  = 100' x 50' x (-3.0)
                  = -15,000 cu ft / 27 = -555.55 cu yds

4. Volume of A_2 = 300' x 100' x [+3.5 - (+1.0)]
                  = 300' x 100' x (+2.5)
                  = +75,000 cu ft / 27 = +2,777.8 cu yds

5. Volume of A_3 = 200' x 150' x [+3.5 - (+3.75)]
                  = 200' x 150' x (-.25)
                  = -7500 cu ft / 27 = -277.77 cu yds

6. Add like signs: -555.55   then subtract: +2777.80
                     -277.77   - 833.32
                     -833.32   +1944.48 cu yds

ANSWER A

ANSWER SHEET

EXCAVATION QUIZ NO. 7

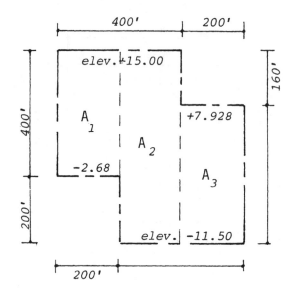

FORMULA FOR VOLUME: $L \times W \times (E_f - E_a)$

1. Solve for missing elevation: From +15.0 to -11.5 there is a total of 26.5'. Because the east side of the property is not in equal increments we must find the rise per foot of run; therefore, 26.5'/600' = .0442' of rise or fall / ft of run. A_1 = 400' x .0442' = 17.68' of fall, 15.0 -17.68 = EL -2.68. A_3 = 160' x .0422' = 7.072' of fall, 15.0 - 7.072 = EL+7.928.

2. Find average elevation:

A_1 = +15.0' $\qquad$ A_2 = +15.0' $\qquad$ A_3 = +7.928'
$\quad$ - 2.68 $\qquad\qquad$ -11.5' $\qquad\qquad$ -11.50 '
$\quad$ + 12.32/2 = +6.16' $\quad$ + 3.5/2 = +1.75' $\quad$ -3.572/2= -1.79

3. Volume of A_1 = 400' x 200' x [-1.5 - (+6.16)]
$\qquad\qquad\qquad$ = 400' x 200' x (-7.66) = -612,800 cu ft

4. Volume of A_2 = 600' x 200' x [-1.5 - (+1.75)]
$\qquad\qquad\qquad$ = 600' x 200' x (-3.25) = -390,000 cu ft

5. Volume of A_3 = 440' x 200' x [-1.5 - (-1.79)]
$\qquad\qquad\qquad$ = 440' x 200' x (+.29) = +25,520 cu ft

6. Add like signs: -   612,800   then subtract: - 1,002,800
$\qquad\qquad\qquad$ -   390,000 $\qquad\qquad\qquad$ +   25,520
$\qquad\qquad\qquad$ - 1,002,800 $\qquad\qquad\qquad$ -   977,280

7. -977,280 cu ft / 27 = -36,195 cu yds to be removed

$\qquad\qquad\qquad\qquad\qquad\qquad\qquad\qquad\qquad\qquad$ ANSWER D

ANSWER SHEET

EXCAVATION PROBLEM NO. 8

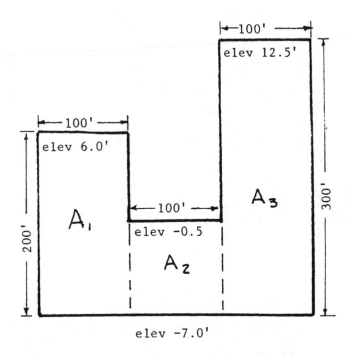

FORMULA FOR VOLUME: $L \times W \times (E_f - E_a)$

1. Solve for missing elevation: From +12.5 to -7.0 there is a total of 19.5'; therefore, 19.5'/3 = 6.5' per 100' of elevation. A_1 = +12.5' - 6.5' = EL 6' and A_2 = 12.5' - 13' = EL -0.5'; therefore, A_2 is 100' in length.

2. Find average elevation:

    A_1 = +6.0 A_2 = -7.0 A_3 = +12.5
         -7.0   -0.5   - 7.0
     -1.0/2 = -.5   -7.5/2 = -3.75   + 5.5/2 = +2.75

3. Volume of A_1 = 200' x 100' x [+2.5 - (-.5)]
                     = 200' x 100' x +3.0 = +60,000 cu ft

4. Volume of A_2 = 100' x 100' x [+2.5 - (-3.75)]
                     = 100' x 100' x (+6.25) = +62,500 cu ft

5. Volume of A_3 = 300' x 100' x [+2.5 - (+2.75)]
                     = 300' x 100' x (-.25) = -7500 cu ft

6. Add like signs: +60,000   then subtract: +122,500
                    +62,500   - 7,500
                    +122,500   +115,000

7. +115,000 cu ft / 27 = +4259.26 cu yds to be brought in.

ANSWER C

FORMWORK

Formwork is a critical part of concrete construction. Ready-mixed concrete delivered to the job site is placed in a fluid consistency and weighs an average of 150 lbs per cu ft. Once the wet concrete is placed into the forms and screeded to the proper level, vibrating or juking is required. Concrete exerts its maximum pressure and weight on the supporting formwork with the initial setting and diminishes within a short period, sometimes less than two hours, after which it reads zero. Therefore, the forms are subjected to maximum stresses for a relatively short period of time.

Lumber for formwork is of two different types. Rough sawed and S4S. Rough sawed lumber is mill cut lumber whose nominal size is the actual size. S4S lumber is finish construction lumber whose nominal size is larger than the actual size.

S4S Lumber

Reduction from actual size:
1" to 1-1/2" reduced by 1/4"
2" to 6" reduced by 1/2"
8" and up reduced by 3/4"

EXAMPLE: 2 x 4 actual size 1-1/2 x 3-1/2

2 x 8 actual size 1-1/2 x 7-1/4

The required book for the examination is FORMWORK FOR CONCRETE, M. K. Hurd. This book explains the functions and formulas for proper concrete design and has comprehensive tables for proper selections of formwork for concrete construction. Examination questions are written primarily to select the proper lumber for formwork construction. See Reference Table, FORMWORK FOR CONCRETE below.

It is strongly recommended that the workbook FORMWORK FOR CONTRACTOR'S EXAMS, 1985, published by the Construction Bookstore, Gainesville, Florida, be used as a part of the study program. Also read and study the entire text PRINCIPLES AND PRACTICES OF HEAVY CONSTRUCTION, R.C. SMITH and C.K. ANDRES

Formwork is used for walls, columns, beams, floor, and roof slabs. Concrete weighs 150 lbs per cu ft.

FIG. 1

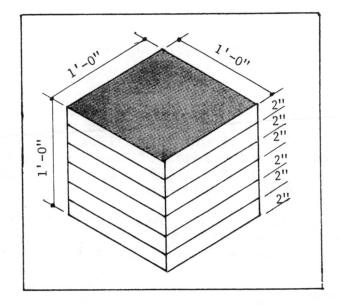

Referring to Fig. 1, one cubic foot of concrete weighs 150 lbs. One sq ft of concrete 2" thick will weigh 25 lbs or for each sq ft, 2 inches thick, concrete will weigh 25 lbs per sq ft. A 4" thick slab will weigh 50 lbs per sq ft. The concrete, steel, etc., that forms the concrete structure is considered the dead load. Any activity such as people, tools, concrete buggy's etc., that it takes to produce the structure are considered the live load. Both dead and live load must be considered when designing formwork.

From PRINCIPLES AND PRACTICES OF HEAVY CONSTRUCTION, R.C. SMITH and C.K. ANDRES, formwork for walls is based on rate of placing concrete in forms and concrete temperature, page 100. Pressure will not exceed 2000 psf. Wall forms are designed using sheathing, wales, and studs, kept together with ties.

Formwork for columns is also based on rate of placing concrete in forms and concrete temperature, pressure will not exceed 3000 psf, see page 100 of PRINCIPLES AND PRACTICES OF HEAVY CONSTRUCTION, SMITH & ANDRES.

In conclusion, it is recommended that the problems in FORMWORK FOR CONTRACTOR'S EXAMS, 1985, Construction Bookstore, be worked out and studied as well as having a good understanding of the text and tables in PRINCIPLES AND PRACTICES OF HEAVY CONSTRUCTION, SMITH & ANDRES.

FORMWORK QUIZ NO. 1

1. GIVEN: A 4-inch thick cast-in-place concrete slab is to be
            poured with a live load of 75 lbs per sq ft. The
            entire weight of the slab and live load is to be
            supported by wood formwork. The joist to be used are
            wood 2 x 6's of S4S f775 psi, E1,100,000 psi, and
            H105 psi, 24 inches on center.

According to PRINCIPLES AND PRACTICES OF HEAVY CONSTRUCTION,
the maximum span of the joist is _____. The weight of
concrete is 150 lbs per cu ft. Do not allow for weight of
forms.

        (A)   4'-0"
        (B)   5'-6"
        (C)   6'-0"
        (D)   6'-9"
        (E)   7'-0"

2. GIVEN: A 12" thick by 12'-0" high by 75'-0" long free
    standing reinforced concrete wall is to be poured.

The formwork is to be constructed of 1" plywood over 4 or
more supports, f = 1930 psi, rolling shear = 80 psi, E =
1,500,000 psi, face grain parallel to span, and S4S lumber of
2 x 6 studs and double 2 x 6 wales. 5000 lb ties are to be
used. Consolidation is by internal vibrator, the concrete
temperature at pouring is 70°F, and the rate of pouring is 7
feet per hr. Studs & wales are spaced over 3 or more spans.

        The following conditions apply:

        Fiber stress   775 psi
        Horizontal shear stress   105 psi
        Modulus of elasticity (E)   1,100,000 psi
        Maximum deflection (d)   length of member/360

Using the tables and procedures of designing forms for walls
in PRINCIPLES AND PRACTICES OF HEAVY CONSTRUCTION, the maximum
spacing of the ties for the wall is_____.

        (A)   12 inches
        (B)   16 inches
        (C)   20 inches
        (D)   23 inches
        (E)   28 inches

FORMWORK QUIZ NO. 2

1. GIVEN: A 6-inch thick cast-in-place concrete slab is to
be poured with a live load of 100 lbs per sq ft.
The entire weight of the slab and live load is to
be supported by wood formwork. The joist to be
used are wood 2 x 6's of S4S f775 psi, E 1,100,000
psi, and H 105 psi Fir, spaced 24" on center.

According to PRINCIPLES AND PRACTICES OF HEAVY CONSTRUCTION
the maximum span of the joist is _____. The weight of
concrete is 150 pounds per cubic foot. Do not allow for
weight of forms.

    (A) 4'-8"
    (B) 5'-2"
    (C) 3'-4"
    (D) 3'-10"
    (E) 4'-2"

2. GIVEN: A 12" thick by 12' high by 100' long free standing
reinforced concrete wall is to be poured.

The formwork is to be constructed of 1" S4S
sheathing over 4 or more supports, f = 800 psi,
H = 120 psi, E = 1,300,000 psi, and S4S lumber of
2" x 6" studs and doubled 2" x 6" wales. 4000 lb
ties are to be used. Consolidation is by internal
vibrator, the concrete temperature at pouring is
60°F, and the rate of pouring is 4 ft per hr.

The following conditions apply:

Fiber stress   775 psi
Horizontal shear stress   105 psi
Modulus of elasticity (E)   1,100,000 psi
Maximum deflection (d)   length of Member/360

Using the tables and procedures of designing forms for walls
in PRINCIPLES AND PRACTICES OF HEAVY CONSTRUCTION the board
feet of the wales is _____.

    (A) less than 1000 bd ft
    (B) between 1000 and 1600 bd ft
    (C) between 1600 and 2200 bd ft
    (D) between 2200 and 2800 bd ft

FORMWORK QUIZ NO. 3

1. GIVEN: A 5" thick concrete slab is poured with a live
            load of 150 pounds per sq ft. The entire weight
            of the slab and live loading is supported by wood
            formwork.

            The formwork joists are constructed of S4S 2" x 8"
            Grade No. 2 f775 psi, E 1,100,000 psi, and H 105 psi
            single span. The concrete weighs 150 lbs per cu ft.

            According to PRINCIPLES AND PRACTICES OF HEAVY
            CONSTRUCTION, a recommended spacing with maximum span
            for joists is _____. Do not allow for weight of
            formwork.

            (A) 24" spacing with maximum span of 5'-0"
            (B) 24" spacing with maximum span of 6'-2"
            (C) 30" spacing with maximum span of 5'-0"
            (D) 30" spacing with maximum span of 6'-2"
            (E) 36" spacing with maximum span of 5'-0"

FORMWORK QUIZ NO. 4

1. GIVEN: The formwork for an 18" x 18" x 12' (high) reinforced concrete column is rated at a maximum pressure of 1178 pounds per sq ft. The temperature of the concrete at pouring is 70°F and the weight of concrete is 150 lbs per cu ft.

   According to PRINCIPLES AND PRACTICES OF HEAVY CONSTRUCTION, the minimum time to pour the column is_____. Select the closest answer.

   (A) 3 ft per hr
   (B) 4 ft per hr
   (C) 5 ft per hr
   (D) 6 ft per hr
   (E) 8 ft per hr

2. What is the column clamp spacing for the third clamp from the bottom if the column pressure is 1800 psf. Use 1-1/4" sheathing over 4 or more supports with, f = 800 psi, H = 120 psi, and E = 1,300,000 psi. Concrete weighs 150 psf. Round spacing down. First clamp is 6" from bottom of column.

   (A)   9"
   (B) 16"
   (C) 27"
   (D) 36"
   (E) 48"

ANSWER SHEET

FORMWORK QUIZ NO. 1

1. Dead Load =   4/12 x 150 = 50 psf
    Live Load =   75 psf
    Total Load =   125 psf x 24/12 = 250 lbs/lin ft

    From page 107: 200 plf = 53"   and   300 plf = 43"

    53" - 43" = 10"/2 = 5" + 43" = 48" = 4'-0"

                                              ANSWER A

2. From page 100: 70°F and 7 fph = 1050 psf (150 x 12 = 1800)
                     1050 psf Governs

    From page 104: spacing of studs is 12" o.c. Pressure on
                    studs is 12/12 x 1050 = 1050 plf.

    From page 106: Spacing of wales is 22" o.c.
                    Pressure on wales is = 22/12 x 12/12 x 1050
                                         = 1,925 PLF

    From page 7-23: Spacing of ties is 23" o.c.

                    $\dfrac{23}{12}$ x $\dfrac{22}{12}$ x   1050 = 3689.28 psi

                                              ANSWER D

ANSWER SHEET

FORMWORK QUIZ NO. 2

1. Dead Load = 6/12 x 150 = 75 psf
    Live Load =   <u>100 psf</u>
    Total Load =   175 psf x 24/12 = 350 plf

    From page 107:
    43" - 37" = 6"/2 = 3" + 37" = 40" = 3'-4"

                                                    ANSWER C

2. From page 100: 60°F and 4 fph = 750 lb pressure
                    (150 x 12 = 1800) 750 lbs governs

    From page 101: Spacing of studs is at 12" o.c. Pressure
                    of studs is 12/12 x 750 = 750 plf
    From page 106: Spacing of wales is 27" - 25" = 2"/2 = 1" + 25
                    26" oc   144/26 = 5.5 spaces = 7 rows of wales

    100' x 2 walls x 2 wales x 7 rows = 2800 lin ft
    2800 lin ft x 6x2/12 = 2800 bd ft

                                                    ANSWER D

ANSWER SHEET

FORMWORK QUIZ NO. 3

1. Dead Load = 5/12 x 150 =   62.5
    Live Load =   150.0
    Total Load =   212.5

    @ 24"/12" x 212.5 = 425   plf
    @ 30"/12" x 212.5 = 562.5 plf
    @ 36"/12" x 212.5 = 637.5 plf

    From page 107

    2 x 8 @ 400 plf =   49
    2 x 8 @ 500 plf = - 44
                        5 #/100 x 25 = 1.25; 49 -1.25 = 47.75 =
                                               4'11-3/4

    2 x 8 @ 500 plf =   44
    2 x 8 @ 600 plf = - 40
                        4 #/100 x 62.5 = 2.5; 44-2.5 =41.5 = 3'5½"

    2 x 8 @ 600 plf =   40
    2 x 8 @ 700 plf = - 37
                        3 #/100 x 37.5 = 1.125; 40 - 1.125 = 38.875
                                                 = 3'2-7/8"

                                        ANSWER A

ANSWER SHEET

FORMWORK QUIZ NO. 4

1. From page 100, Table 6-3, at 70°F and 1178 lbs
   per sq ft the rate of filling forms is 8 hrs.

                                                    ANSWER E

2. From page 101: Maximum pressure on second clamp is 1800 psf
                   (6/12 x 150) = 1725 psf. Interpolating:

         1600 = 11"   -   1800 = 10"

         $1725 - 1600 = \dfrac{125}{X}$: $\dfrac{200}{1}$ = .625"

         10" + .625" = 10.625" = 10"

Maximum Pressure for second clamp = 6" + 10" = 16"

1800 - (16"/12" x 150) = 1600 psi = 11" spacing

6" + 10" + 11" = 27"

                                                    ANSWER C

PLAN READING

Plan reading and comprehension of the specifications is most essential, primarily for the first day of the examination. The test writers specifically note that the plans are "not for construction", which indicates that certain information could be deliberately omitted causing the candidate to search for or find other mathematical means of finding information. Some dimensions may have to be found by means of calculation.

The notes of the plan drawings and the specifications pertaining to the plans will supercede the actual plan drawings. It is important to be aware of the information contained in both the notes and the specifications and watch that the information does not conflict. Both notes and specs should be read thoroughly and highlighted.

A sheet of symbols and abbreviations will accompany the specs which will show section directions. Should more information be required on symbols and abbreviations, refer to WALKER'S ESTIMATOR'S REFERENCE BOOK, 3.7-3.20.

When searching for information from the plans always check the plan views, elevation views, and details. The test writers have a tendency to hide information in certain areas and the information must be found elsewhere. Many times a question will be asked regarding information at one location on the plans and in order to determine a dimension at that location it is necessary to go to another location, total a line of dimensions and subtract from a dimension at the required location. For example, there are times when, in order to find an area at the top of a particular plan sheet it is necessary to go to some other sheet of the plans and total a series of dimensions at the bottom of the plan.

Be aware of elevation information. The symbol (⊕) with a number associating it, or the letters EL with a number indicates a given elevation. Elevations vary in the State of Florida from coastal areas to inland areas with elevations ranging from a few feet above sea level (sea level = 0.00') to inland elevations of +100.00'.

Elevations will change from bench mark or land elevations to a fixed location elevation equaling 0.00'. For example, the finish first floor elevation could have a reference from bench mark of 98.50'. It will then be denoted that elevation 98.50' equals 0.00' and foundation dimensions will be shown as negative numbers. If the finish first floor elevation is shown as 98.50' = 0.00' and the top of the basement footers are shown at -5.00' the footers would actually be at an elevation of 93.50'.

If the candidate does not have a strong knowledge of plan reading and specifications, it is recommended that Chapter 3, page 3.1 of WALKER'S ESTIMATOR'S REFERENCE BOOK be read in its entirety. The best approach to plan reading for the State Exam is to practice from old test plans with answers, rather than attempt to learn plan reading from actual job plans.

104

ELEVATION IDENTIFICATION

From the elevations given below, place the letter of the correct elevation in the proper blank to match the building.

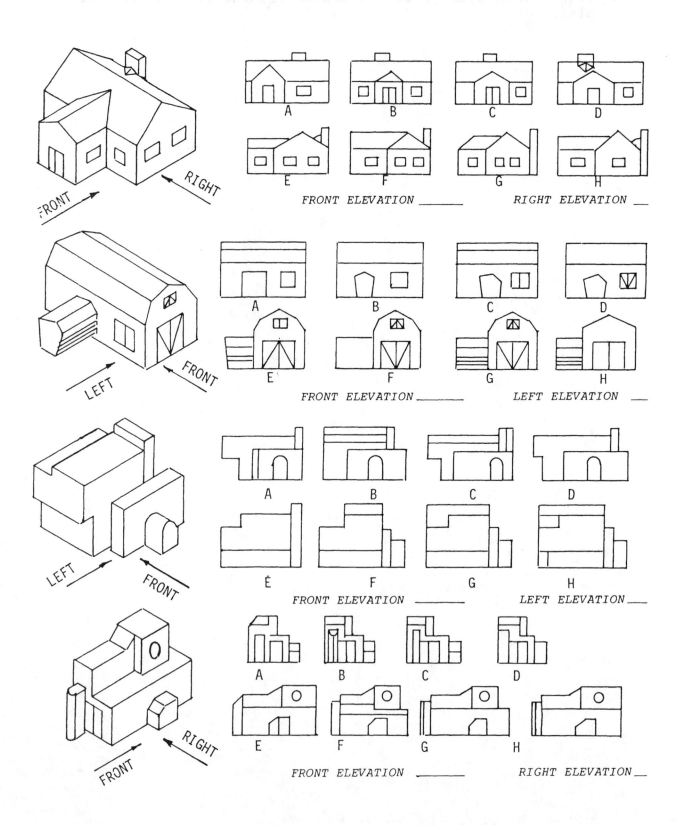

FRONT ELEVATION _____   *RIGHT ELEVATION* __

FRONT ELEVATION _____   *LEFT ELEVATION* __

FRONT ELEVATION _____   *LEFT ELEVATION* __

FRONT ELEVATION _____   *RIGHT ELEVATION* __

105

ANSWER SHEET

ELEVATION IDENTIFICATION

<u>Note</u>: As shown below, shade one elevation with a yellow highlighter, which will make both elevations much easier to identify.

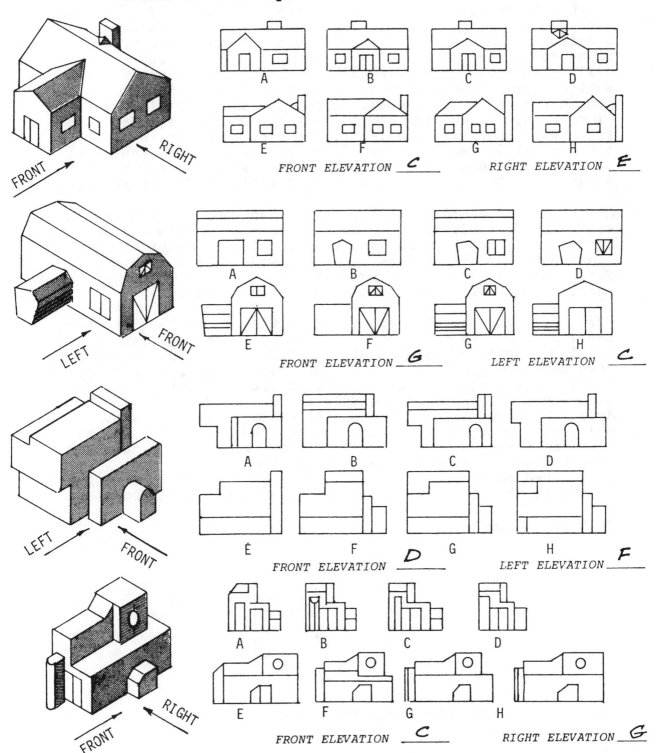

FRONT ELEVATION __C__   RIGHT ELEVATION __E__

FRONT ELEVATION __G__   LEFT ELEVATION __C__

FRONT ELEVATION __D__   LEFT ELEVATION __F__

FRONT ELEVATION __C__   RIGHT ELEVATION __G__

SITE PLANS

Whenever a set of construction plans are created by the
Architect, a site plan must be included. The site plan
includes the legal boundaries of the site, as well as the
land elevation, building location, parking areas, etc.

A typical question asked on the State Examination is, to find
the area of the site within the property lines. This is just
a matter of mathematical calculations. The biggest problem
is searching for unknown dimensions. In many cases a length
will be left out of a triangle and the Pythagorean's Theorem
must be applied (See Formula on page 39). At other times the
configuration of the property lines can be misleading as to
dimensions, and an area can be missed. There are times that
the answer band for the area of the site is given in sq ft,
although it is usually given in acres; therefore, the total
sq ft must be converted to acres. From WALKER'S ESTIMATOR'S
REFERENCE BOOK, p 26.2. **ONE ACRE IS EQUAL TO 43,560 SQ FT.**

Questions are asked regarding sodded and seeded areas, which
are numbered 1, 2, 3, etc. Be careful that sodded and not
seeded is calculated and visa versa. Also be cautious in
reading the specifications regarding sodding and seeding.
Watch that the dimensions are taken to the proper side of the
curbing when calculating sodded or seeded areas.

Curbing is always referred to in the test questions. Either
the length of the curb, or volume of concrete required in all
curbing is asked. Curbing will include arc's and all turns.
(See Math Section p 40 for Circumference of quarter Circle.)
To find the length of curbing always use a center line
dimension, as curbing is usually 18" wide. Use caution when
determining the dimension from the radius arrow of curbing as
there are times the arrow is shown to the inner side of the
curb and times it is shown to the outer side of the curb.
When the square area of the section thru the curb is
calculated make sure that the product is converted to sq ft,
as the plans will show the detail in both feet and inches.

EXAMPLE: <u>Given</u>:

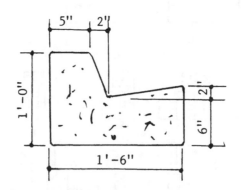

Find the cross section of the curb.

SOLUTION:

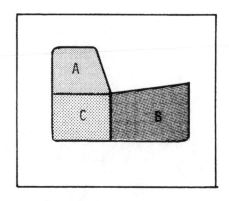

<u>Using Averaging</u>:

Section "A" =   5" + 7"/2   x   6"   =   36 sq in

Section "B" =   6" + 8"/2   x   11" =   77 sq in

Section "C" =   7"   x   6"   =   42 sq in

36 + 77 + 42 = 155 sq in. / 144 sq in. = 1.077 sq ft

<u>Note</u>: From WALKER'S ESTIMATOR'S REFERENCE BOOK, page 26.2.
          1 sq ft = 144 sq in.

It is recommended that colored pencils or highlighters be
used to mark various parts of the site plan for easier
calculation. Be careful that the right dimensions are used
for calculations. Because of the size of the plans, markings
can be misleading and the wrong dimension selected.

PROPERTY AREA PROBLEM NO. 1

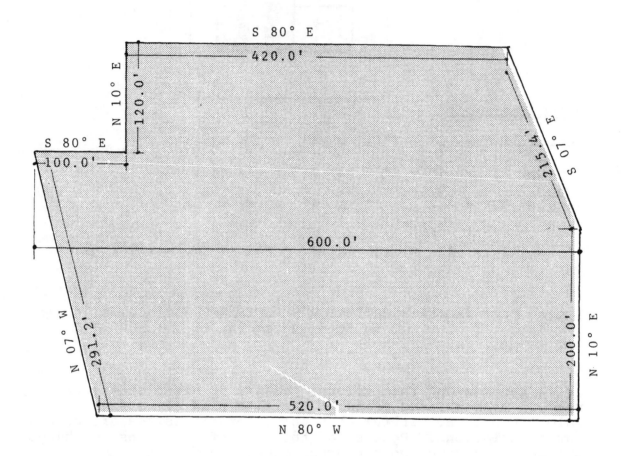

FIND THE AREA WITHIN THE PROPERTY LINES.

    A. less than 4.0 acres
    B. between 4.0 and 4.5 acres
    C. between 4.5 and 5.0 acres
    D. between 5.0 and 5.5 acres
    E. between 5.5 and 6.0 acres

PROPERTY AREA PROBLEM NO. 2

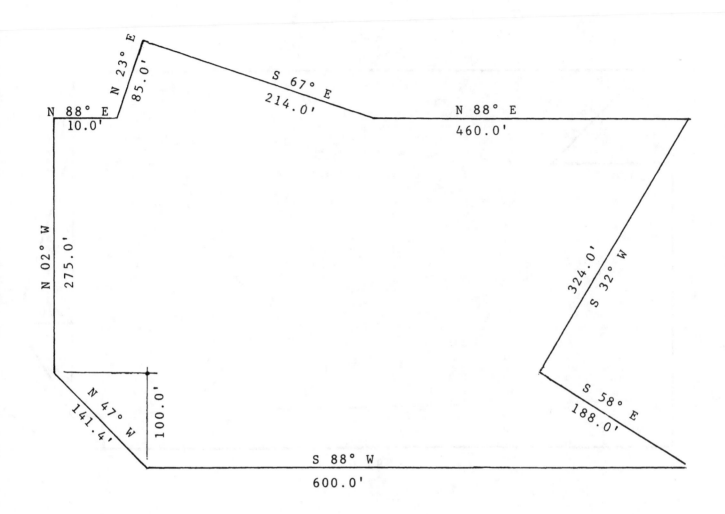

FIND THE AREA WITHIN THE PROPERTY LINES.

- A. between 4.75 and 5.2 acres
- B. between 5.2 and 5.6 acres
- C. between 5.6 and 6.0 acres
- D. between 6.0 and 6.5 acres
- E. greater than 6.5 acres

PROPERTY AREA PROBLEM NO. 3

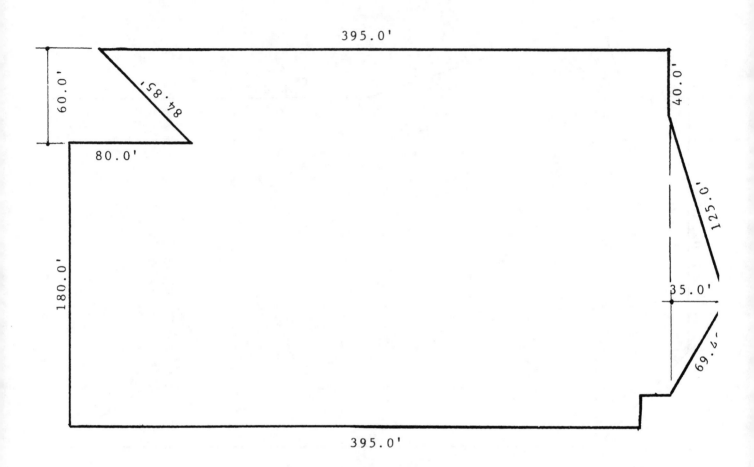

FIND THE AREA WITHIN THE PROPERTY LINES.

- A. less than 2.0 acres
- B. between 2.0 and 2.5 acres
- C. between 2.5 and 3.0 acres
- D. between 3.0 and 3.5 acres
- E. greater than 3.5 acres

PROPERTY AREA PROBLEM NO. 4

Find the area within the property
lines.

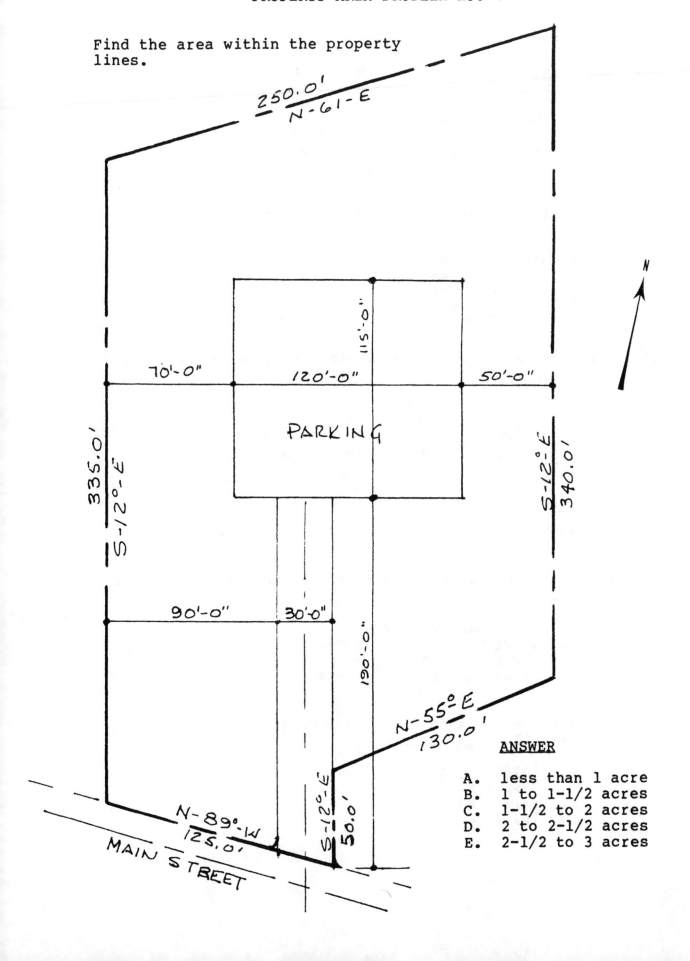

ANSWER

A. less than 1 acre
B. 1 to 1-1/2 acres
C. 1-1/2 to 2 acres
D. 2 to 2-1/2 acres
E. 2-1/2 to 3 acres

SITE PLAN QUIZ

(Use Site Plan on following page)

1. The total area of the site within the property lines is _____.

    A. between 1.00 and 1.99 acres
    B. between 2.00 and 2.09 acres
    C. between 2.10 and 2.19 acres
    D. between 2.20 and 2.29 acres

2. The angle I located in the S.E. corner of the site is _____.

    A. 104° 54' 28"
    B. 105° 23' 58"
    C. 105° 46' 35"
    D. 106° 22' 11"

3. The total area of paving within the property lines is _____.

    A. less than 16,500 sq ft
    B. between 16,500 and 17,500 sq ft
    C. between 17,500 and 18,500 sq ft
    D. between 18,500 and 19,500 sq ft

4. The total length of concrete curb and gutter is _____.

    A. 746'   B. 779'   C. 813'   D. 847'

5. The total net yards of concrete in the curb and gutter is _____.

    A. less than 32 cu yds
    B. between 32 and 34 cu yds
    C. between 34 and 38 cu yds
    C. between 38 and 42 cu yds

6. The total area of seeded area located West of the office building is _____.

    A. 14,900 sq ft
    B. 15,250 sq ft
    C. 15,600 sq ft
    D. 16,320 sq ft

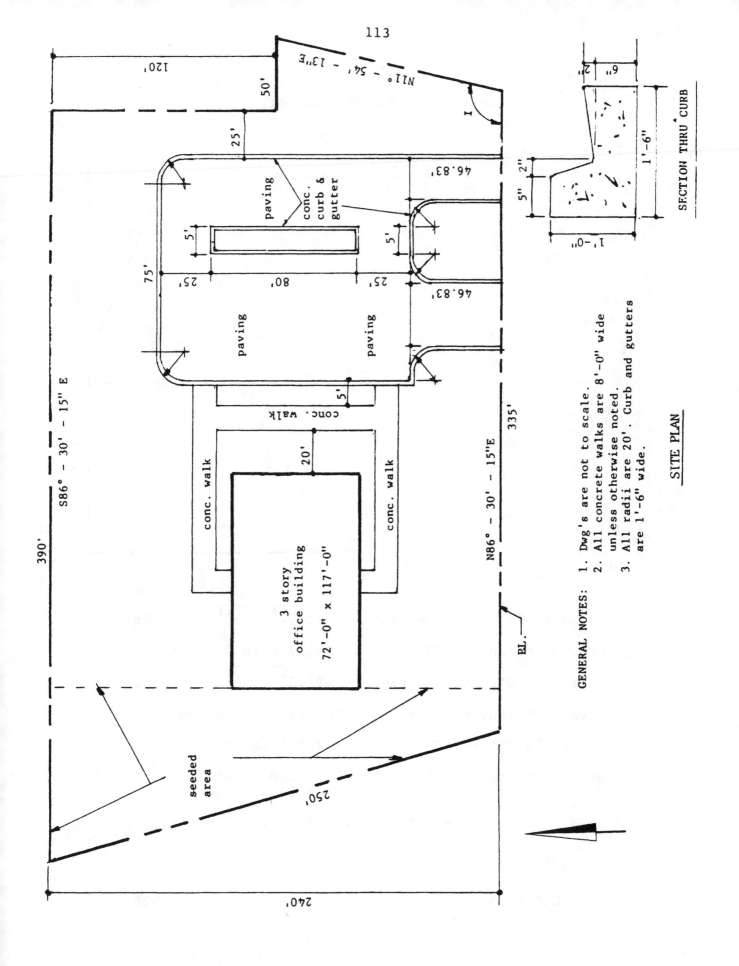

SECTION THRU CURB

SITE PLAN

GENERAL NOTES: 1. Dwg's are not to scale.
                 2. All concrete walks are 8'-0" wide
                    unless otherwise noted.
                 3. All radii are 20'. Curb and gutters
                    are 1'-6" wide.

N11° - 54' - 13" E

S86° - 30' - 15" E

N86° - 30' - 15" E

390'

250'

240'

120'

50'

25'

75'

25'

80'

25'

46.83'

46.83'

335'

20'

5'

5'

5'

5'

3 story
office building
72'-0" x 117'-0"

conc. walk

conc. walk

conc. walk

paving

paving

paving

conc.
curb &
gutter

seeded
area

P.L.

1'-0"

5"

2"

2"

6"

1'-6"

ANSWER SHEET

PROPERTY AREA PROBLEM NO. 1

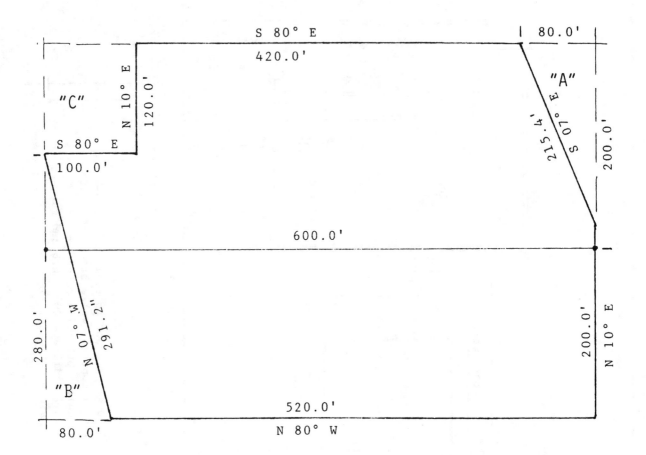

1. Triangle "A", unknown side = $\sqrt{215.4^2 - 80^2}$ = 200 ft.

2. Area of triangle "A" = 80' x 200'/2 = 8000 sq ft.

3. Triangle "B", unknown side = $\sqrt{291.2^2 - 80^2}$ = 280 ft.

4. Area of triangle "B" = 80' x 280'/2 = 11,200 sq ft.

5. Rectangle "C" = 100' x 120' = 12,000 sq ft.

6. Gross area of overall rectangle = 600' x 400'
    = 240,000 sq ft

7. [(240,000 sq ft) - (8000 + 11,200 + 12,000 sq ft)]

    = $\dfrac{208,800}{43,560}$ = 4.79 acres

ONE ACRE = 43,560 SQ FT   ANSWER C

ANSWER SHEET

PROPERTY AREA PROBLEM NO. 2

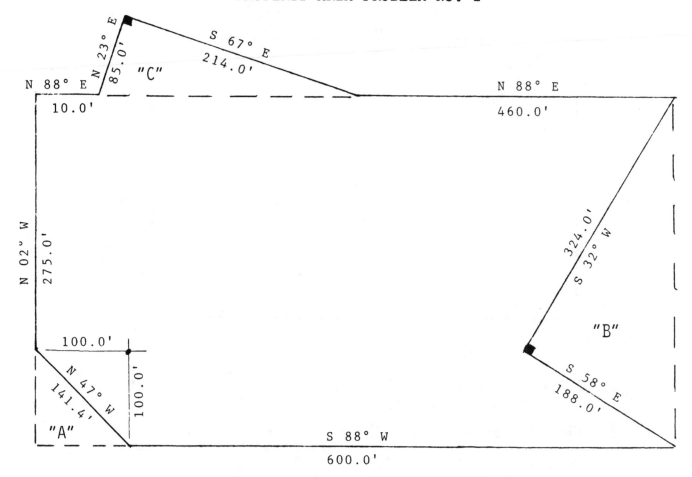

1. Unknown dimension of "A" = $\sqrt{141.4^2 - 100^2}$ = 100 ft

2. Area of triangle "A" = 100' x 100'/2 = 5000 sq ft.

3. Angle "B" is a right triangle (by plotting property lines 58 + 32 = 90°). Area = 324' x 188'/2 = 30,456 sq ft.

4. Angle "C" is a right triangle (by plotting property lines 67 + 23 = 90°). Area = 85' x 214'/2 = 9095 sq ft.

5. Gross area of large rectangle = 700' x 375'
    = 262,500 sq ft

6. 262,500 - 30,456 - 5000 + 9095

    = $\dfrac{236,139}{43,560}$ = 5.42 Acres.

    ONE ACRE = 43,560 SQ FT   ANSWER B

ANSWER SHEET

PROPERTY AREA PROBLEM NO. 3

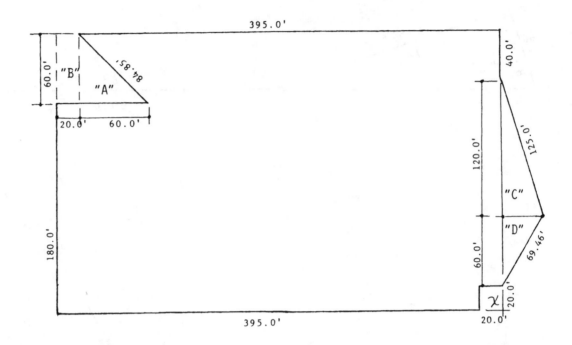

1. Unknown dimension of "A" = $\sqrt{84.85^2 - 60^2}$ = 60 ft
   Area of "A" = 60' x 60'/2 = 1800 sq ft

2. Area of "B" = 60' x 20' = 1200 sq ft.

3. Unknown dimension of "C" = $\sqrt{125^2 - 35^2}$ = 120 ft
   Area of "C" = 120' x 35'/2 = 2100 sq ft

4. Unknown dimension of "D" = $\sqrt{69.46^2 - 35^2}$ = 60 ft
   Area of "D" = 60' x 35'/2 = 1050 sq ft

5. Area of "X" = 20' x 20' = 400 sq ft

6. Gross area: = 415'x 240' =   99,600 sq ft
        plus: Area "C"   =   2,100 sq ft
                  Area "D"   =   1,050 sq ft
                                   102,750 sq ft

7. Area not in property area: Area "A" =   1,800 sq ft
                               Area "B" =   1,200 sq ft
                               Area "X" =   400 sq ft
                                           3,400 sq ft

8. Net Area = 102,750 - 3,400 = $\dfrac{99,350}{43,560}$ = 2.28 acres

   <u>ONE ACRE = 43,560 SQ FT</u>   ANSWER B

ANSWER SHEET

PROPERTY AREA PROBLEM NO. 4

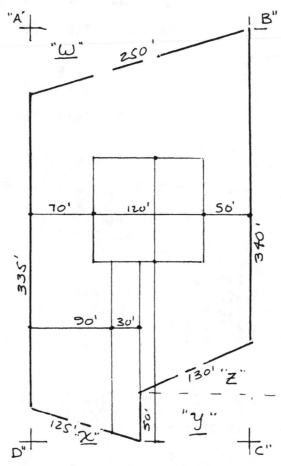

1. S-12°-E are parallel lines making A-B-C-D a rectangle.

2. Unknown dimension of "W" = $\sqrt{250 - 240}$ = 70'
    Area of "W"   = 70' x 240'/2 = 8400 sq ft

3. Unknown dimension of "X" = $\sqrt{125 - 120}$ = 35'
    Area of "X"   = 35' x 120'/2 = 2100 sq ft

4. Area of "Y" = 120'(240'-90'-30') x 50' = 6000 sq ft

5. Area of "Z" = 50'(35'+335'+70'-340'-50') x 120'/2
                                        = 3000 sq ft

6. Total of W,X,Y,Z: 8400 + 2100 + 6000 + 3000= 19,500 sq ft

7. Gross Area of Rectangle = 440' x 240' = 105,600 sq ft

8. Net Area = 105,600 - 19,500 = $\dfrac{86,100}{43,560}$ = 1.98 acres

   ONE ACRE = 43,560 SQ FT   ANSWER C

118

ANSWER SHEET - SITE PLAN QUIZ

1. (B)

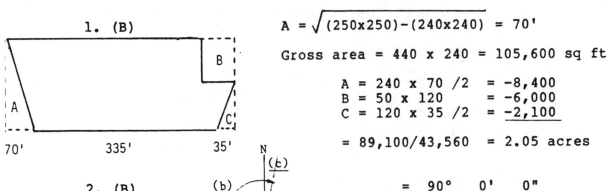

70'   335'   35'

$$A = \sqrt{(250\times250)-(240\times240)} = 70'$$

Gross area = 440 x 240 = 105,600 sq ft

A = 240 x 70 /2	= -8,400
B = 50 x 120	= -6,000
C = 120 x 35 /2	= -2,100

= 89,100/43,560 = 2.05 acres

2. (B)

```
        = 90°   0'   0"
       -86°  30'  15"
(a) =    3°  29'  45"
+(b) = 90°   0'   0"
+(c) = 11°  54'  13"
Angle I = 105°  23'  58"
```

3. (A)

	130' x 112'	= 14,560 sq ft
plus -	46.83' x (112' - 65')	= 2,201
less -	.215 x 18.5' x 18.5' x 2	= -147
plus -	.215 x 20' x 20' x 3	= 258
less -	80' x 5'	= -400
		16,472 sq ft

4. (A)

Radii = (40' - 1.5)' = 38.5 x 3.14/4	
= 30.223 x 5 + 5' =	156.115'
(46.83' - 19.25') = 27.58 x 3 =	82.740'
46.83 x 1 =	46.830'
(130' - 19.25') x 2 =	221.500'
75' x 1 =	75.000'
planter = 78.5 x 2 =	157.000'
planter = 3.5 x 2 =	7.000'
	746.185'

5. (A)

Gross area =	1.5 x 1.0	=	1.5 sq ft
less:	A = .333 x .917	=	.305
	B = .167 x .5/2	=	.042
	C = .167 x .917/2 =		.077
			1.076 sq ft

1.076 x 741.2 = 797.53/27 = 29.5 cu yds.

6. (D) From right to left (25 + 112 + 5 + 8 + 20 + 117)= 287
390' - 287' = 103' at top, and
(335' + 35') = 370' - (287' + 50') = 33' at bottom.
Averaging 103' + 33'/ 2 = 68' x 240' = 16,320 sq ft.

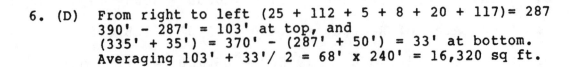

PROPERTY LINES AND PROPERTY ANGLES

Property lines and property angles are a part of surveying
which is a branch of applied mathematics dealing with the
science of measuring land. The unit of measure being the
surveyor's chain, with 80 chains equal to one mile. A survey
is a legal description of a plot. The property lines are the
recorded boundaries of the plot, and the property angles are
the directions of enclosure. The property line system is a
system established in colonial days called the Metes and
Bounds System, which describes a plot of land by starting at
a designated boundary point, known as the point of beginning
(POB) and then proceeding to the directions and distance of
the property lines until the property perimeter has been
traced back to the point of beginning. Distances are defined
in feet and the decimal part of a foot. Orientation is
defined by compass point.

The compass and surveyors instrument are designed like a
circle for a total of 360°; the degrees are further divided
into 60 minutes, and the minutes into 60 seconds. Therefore,
90°-00'-00" may also be written 89°-59'-60". If 62°-46'-51"
were to be subtracted from 90°-00'-00", borrow 1 minute from
90°, changing it to 89°-60'-00", then borrow 60 seconds from
the 60 minutes changing it to 89°-59'-60", then subtracting:

$$
\begin{array}{rrr}
89° & 59' & 60" \\
-62° & 46' & 51" \\
\hline
27° & 13' & 9" \\
\end{array}
$$

In the same respect, by addition:

$$
\begin{array}{rrr}
89° & 59' & 60" \\
62° & 46' & 51" \\
\hline
151° & 105' & 111" \\
\end{array}
$$

Now there are over 60 minutes and 60 seconds; therefore,
subtract 60 in each case and move one to the left:

$$
\begin{array}{rrr}
151° & 105' & 111" \\
 & & 60" \\
\hline
151° & 106' & 51" \\
 & 60' & \\
\hline
152° & 46' & 51" \\
\end{array}
$$

In the Metes and Bounds Method of surveying, the procedure is
to go either to the north or to the south; therefore, all
directions are from the north to the east or west, or from
the south to the east or west, and the distance is never more
than 90°-00'-00" in any direction. Directions are written
(N to E) or (N to W) or (S to E) or (S to W).

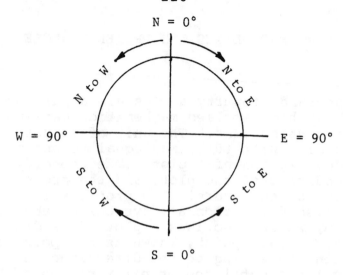

FIG. 1

To plot a site the surveyor must always be given a point of beginning (POB), a direction from the POB, and an orientation.

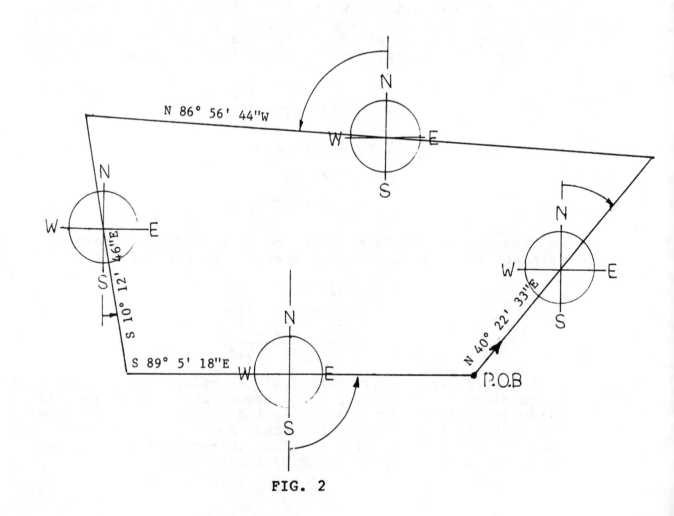

FIG. 2

Orientation co-ordinates will never be over 90°-00'-00". The property angle may be anywhere from 1 degree to 359 degrees.

In past examinations the test writers have always asked one or two questions regarding property angles. They will want to know the angle formed between two property lines.

To determine the angle, draw a diagram to scale with four 90° quadrants using a protractor.

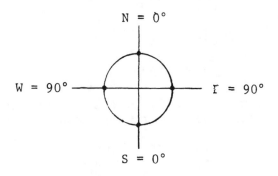

FIG. 3

From Fig. 2, plot the angle formed by the northernmost property line and the easternmost line, keeping in mind that the protractors flat side is always placed on the N-S line,

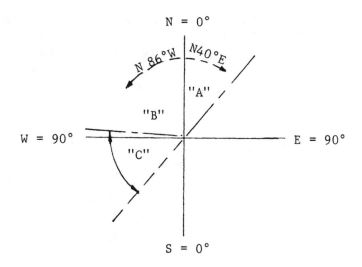

FIG. 4

The angle called for in Fig. 2 is an acute angle (less than 90°). The angle plotted in Fig. 4 is an obtuse angle (greater than 90°). All four quadrants are symetrical (equal size). A straight line is equal to 180° angle; therefore, by intersecting into the SW quadrant, a 180° angle (straight line) is formed. The angle asked for in Fig. 2 is angle "C" in Fig. 4. To find angle "C", first add angles "A" and "B":

```
 40°   22'   33"
 86°   56'   44"
126°   78'   77"
             -60"
126°   79'   17"
       60'
127°   19'   17"
```

127°-19'-17" = A and B; A, B, and C = 180°-00'-00" = 179°-59'-60", subtracting (A and B) from the total;

```
179°   59'   60"
127°   19'   17"
 52°   40'   43"
```

which is angle "C".

In order to develop a property site there must be a point of beginning (POB), a direction of procedure from that point of beginning and an orientation. The problems asked on the examination plans show neither a POB nor a direction. Therefore, when an angle is plotted, it will either be the angle desired or the reciprocal (opposite) of the angle desired.

Whenever an angle is asked for, always extend the lines to form four quadrants which gives a visual view of the opposite angle equaling 180°. Opposite angles are always equal.

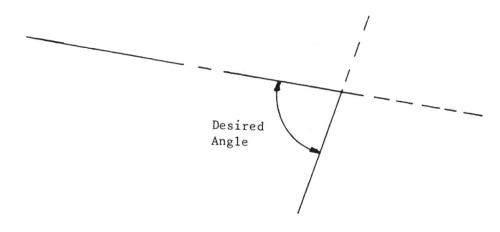

FIG. 5

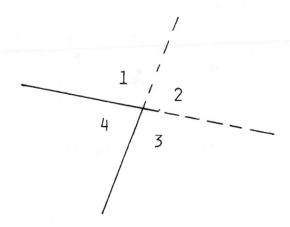

FIG. 6

In Fig. 6, angles 1 and 3 are equal, and angles 2 and 4 are equal.

The sure way to determine a property angle is to draw a scale drawing of four quadrants as shown in Fig. 3, then plot the coordinates as shown in Fig. 4. Keep in mind that whenever an angle in one quadrant is projected into the opposite quadrant, it will have the same value.

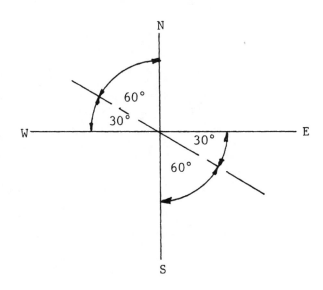

FIG. 7

The State Exam plans note that "the plans are not to scale" although they are projected from scale drawing; therefore, the protractor may be placed on the angle asked for and an approximate answer established.

In most cases the State's answer band will have two or three answers close enough that calculations must be computed to determine the right answer. In very few cases will the protractor, when layed on the angle, give the exact answer.

PRACTICE PROBLEMS

Add the following co-ordinates:

1.	43°	41'	11"	plus	37°	33'	53"
2.	72°	30'	30"	plus	82°	53'	43"
3.	12°	12'	10"	plus	6°	47'	49"
4.	162°	52'	43"	plus	197°	07'	17"
5.	82°	18'	22"	plus	97°	41'	38"

Subtract the following co-ordinates:

6.	33°	33'	33"	from	86°	31'	23"
7.	60°	52'	34"	from	122°	15'	16"
8.	26°	22'	35"	from	27°	53'	52"
9.	130°	10'	12"	from	342°	59'	59"
10.	104°	4'	9"	from	180°	00'	00"

PROPERTY ANGLES

What is the angle between the property lines?

1.

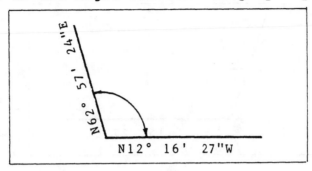

2.

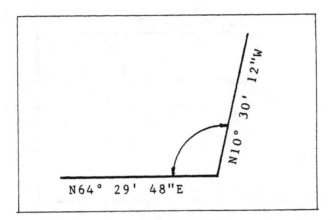

3.

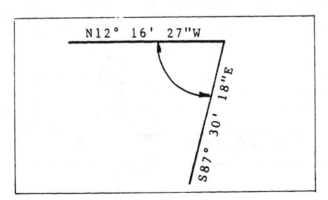

4.

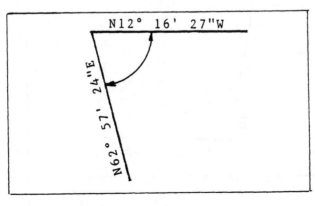

PROPERTY ANGLES

What is the angle between the property lines?

5.

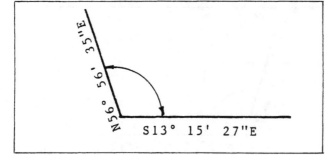

6.

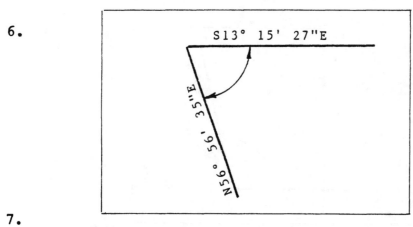

7.

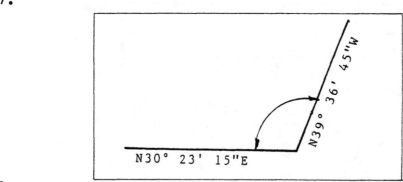

8.

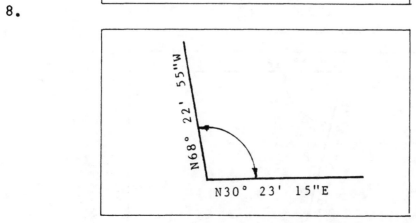

127

ANSWER SHEET

PRACTICE PROBLEMS

1.
```
  43°  41'  11"
  37°  33'  53"
  80°  74'  64"

  81°  15'  04"
```

6.
```
  86°  31'  23"
 -33°  33'  33"
  52°  57'  50"
```

2.
```
  72°  30'  30"
  82°  53'  43"
 154°  83'  73"

 155°  24'  13"
```

7.
```
 122°  15'  16"
 -60°  22'  34"
  61°  22'  42"
```

3.
```
  12°  12'  10"
   6°  47'  49"
  18°  59'  59"
```

8.
```
  27°  53'  52"
 -26°  22'  35"
   1°  31'  17"
```

4.
```
 162°  52'  43"
 197°  07'  17"
 359°  59'  60"

 360°  00'  00"
```

9.
```
 342°  59'  59"
-130°  10'  12"
 212°  49'  47"
```

5.
```
  82°  18'  22"
  97°  41'  38"
 179°  59'  60"

 180°  00'  00"
```

10.
```
 180°  00'  00"
-104°  04'  09"
  75°  55'  51"
```

ANSWER SHEET

PROPERTY ANGLES

1.

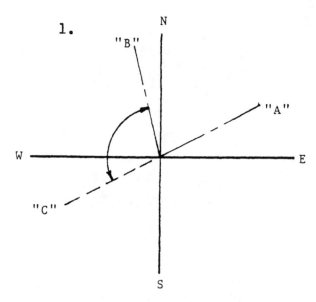

"A" 62° 57' 24"
"B" 12° 16' 27"
   75° 13' 51"

"AC" 180° 00' 00"
"AB" -75° 13' 51"
"BC" 104° 46' 09"

2.

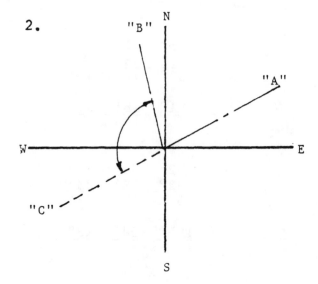

"A" 64° 29' 48"
"B" 10° 30' 12"
   75° 00' 00"

"AC" 180° 00' 00"
"AB" 75° 00' 00"
"BC" 105° 00' 00"

3.

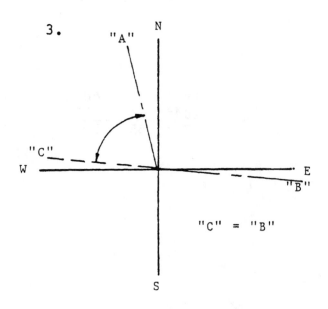

"C" = "B"

"C" 87° 30' 18"
"A" 12° 16' 27"
   75° 13' 51"

4.

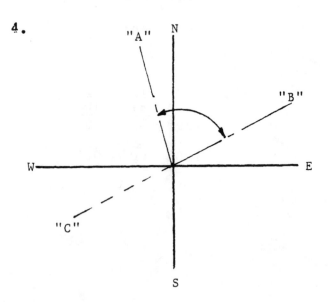

"A" 12° 16' 27"
"B" 62° 57' 24"
"AB" 75° 13' 51"

ANSWER SHEET

PROPERTY ANGLES

5.

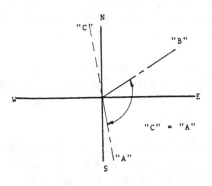

```
"B"   56° 56' 35"
"C"   13° 15' 27"
      70° 12' 02"

"AC"  180° 00' 00"
    -  70° 12' 02"
"BA"  109° 47' 58"
```

6.

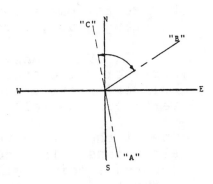

```
"C"   13° 15' 27"
"B"   56° 56' 35"
"CB"  70° 12' 02"
```

7.

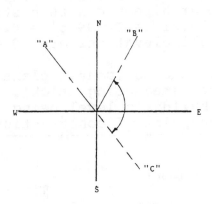

```
"A"   30° 23' 15"
"B"   39° 36' 45"
      70° 00' 00"

"AC"  180° 00' 00"
"AB" -  70° 00' 00"
"BC"  110° 00' 00"
```

8.

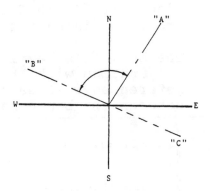

```
"A"   68° 22' 55"
"B"   30° 23' 15"
"AB"  98° 46' 10"
```

THE TRANSIT - LEVELING INSTRUMENT

There are two types of leveling instruments known as the builder's level and the transit-level. (See ARCHITECTURAL AND BUILDING TRADES DICTIONARY, Third Edition, R.E. Putnam, A.E. Carlson, pages 73 and 267, and page 459.)

    1. The builder's level consists of a sighting glass with a leveling device mounted on a tripod. When leveled, a horizontal plane can be read at various distances through the sighting glass, through 360°.

    2. The transit-level is an instrument used by surveyor's and contractors. The principle is the same as the builder's level, except that the transit rotates to read vertically as well as horizontally.

One of the first steps in construction is to establish elevations. In order to establish job site elevations a reference point called a bench mark must be given. Bench marks are elevations above sea level established by surveyors, (sea level being elevation -0-). Bench marks can be located by a marker in the crown of the road, concrete markers located in the ground by the surveyor, or other permanently fixed locations.

Most plans identify bench marks by the symbol ⊕ , or the letters BM, or EL. In the case of the State Examination plans the bench mark reference may also be found in the notes or specifications. Many times the surveys reference bench mark is transposed to the building site and referenced as -0-. For example; The reference bench mark at the crown of the road reads EL 97.80; the architect can then state on the plan that finish first floor level is EL 99.5' = 0.0'. Then from the finish first floor, all elevations are +0 or -0.

The builder's level is used in the horizontal plane only in conjunction with the stadia rod (measuring stick). Most references will describe the builder's level being set between the bench mark and the point of construction.

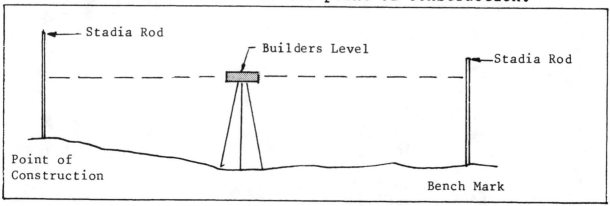

FIG. 1

This is fine for job conditions especially where the distance is great between points.

For calculation purposes pertaining to exam problems it is recommended that the builder's level be the starting point and all other markings continue through the level line.

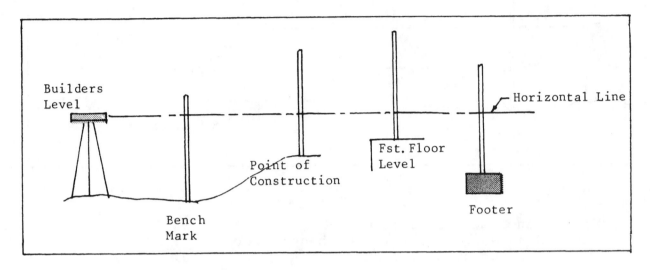

FIG. 2

The first step is to sight the reading on the stadia rod at the bench mark, then all other readings on the stadia rod can be added or subtracted from the stadia rod reading at the bench mark to determine their elevations.

EXAMPLE: Given: The bench mark at the crown of the road reads EL 94.50', the stadia rod placed on the bench mark reads 7.25' through the sighting glass of the builder's level. When the stadia rod is placed at the point of construction to dig the footer it reads 3.50'. The finish first floor is EL 100.0' = -0-. The top of the footer is at EL -5.0 and the footer is 18 inches deep. How far down does the builder have to dig at the point of construction to establish the bottom of the footer?

SOLUTION: All points are referenced from the bench mark.

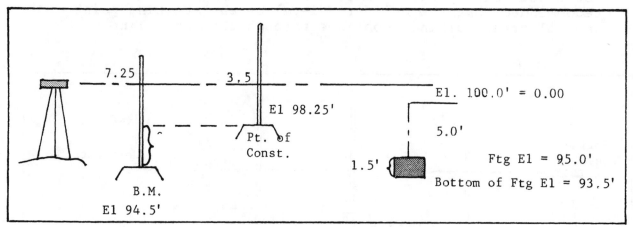

FIG. 3

Bench mark elevation   94.50'
Stadia rod elevation at bench mark   7.25'
Instrument elevation   = 101.75'

Instrument elevation   101.75'
Point of construction   3.50'
Point of construction elevation   = 98.25'

Point of construction elevation   98.25'
Bottom of footing elevation   93.50'
Dig   = 4.75'

TRANSIT PROBLEMS QUIZ NO. 1

1. The bench mark on Monroe Street reads +92.0'. The plans show finish floor level at +100.0'. The transit is set between the building and bench mark. The stadia rod reading at the bench mark is 7.25'. The stadia rod reading at the top of the footer is 2.5' and the footer is 18" deep.

   What is the distance between the finish floor and the bottom of the footer?

         A. 1.75'
         B. 2.75'
         C. 3.75
         D. 4.75'

2. You are setting your batterboards at elevation 12.0'. If the bench mark is 9.42' and you read 6.27' on the stadia rod placed on the bench mark, what would you read on the stadia rod when the batterboards were at the proper elevation?

         A. 1.09
         B. 2.56
         C. 3.69
         D. not enough information given

3. The bench mark shown in the lower right hand corner of the plans is at EL 93.50'. The column schedule shows the top of the footer for column C-2 to be at elevation -5.00', and shows the footer as 6-0' x 6-0' x 2-0' deep. Finish first floor is EL 100.00' = 0.00'. When placing the stadia rod at the bench mark the rod reads 7.75'. When placing the stadia rod at location on column C-2 the reading is 2.5'. How deep will the contractor dig to establish this footer?

         A. 4.74'
         B. 5.25'
         C. 5.75'
         D. 5.95'

TRANSIT PROBLEMS QUIZ NO. 2

1. When constructing a house the bench mark reading is at
   EL 6.25'. The stadia rod reading is 5.5'.
   Batterboards are set at EL 9.2' F.F. and the bottom of
   the footer is 2.25' below the finish grade. What is the
   stadia rod reading at the bottom of the footer?

   A. 3.25'
   B. 4.80'
   C. 5.50'
   D. 6.25'

2. A store building must have a wainscot at 4' 2". The
   ceiling height is 12.0'. At a given working point at
   floor level the rule will read 5' 4-1/2". What is the
   reading on the rule to snap a line for the wainscot?

   A. 1'-½"
   B. 1'-0"
   C. 1'-2½"
   D. 1'-5¼"

3. A 60' ramp is being built with a total vertical rise of
   6'-0" the elevation at the bottom of the ramp is 10.5'.
   The stadia rod is reading 5.167' when placed at the
   bottom of the ramp. An expansion joint must be placed at
   the 1/3 point from the bottom of the ramp. What would be
   the stadia rod reading at this point? Select the closest
   answer.

   A. 1'-9"
   B. 2'-0"
   C. 2'-8"
   D. 3'-2"

TRANSIT PROBLEM QUIZ NO. 3

1. The bench mark on Monroe Street is 127.52'. The transit
    is 5' 2" off the ground. The stadia rod at the bench
    mark is reading 8.55'. The stadia rod at the location of
    footer Mark 2 is reading 14.25'. The first floor finish
    slab is at EL 131.36 = 0.00'. The footer is at EL -8.0'
    and the footer is 2'8" deep. How far must you go down at
    Mark 2 to establish the bottom of the footer?

2.

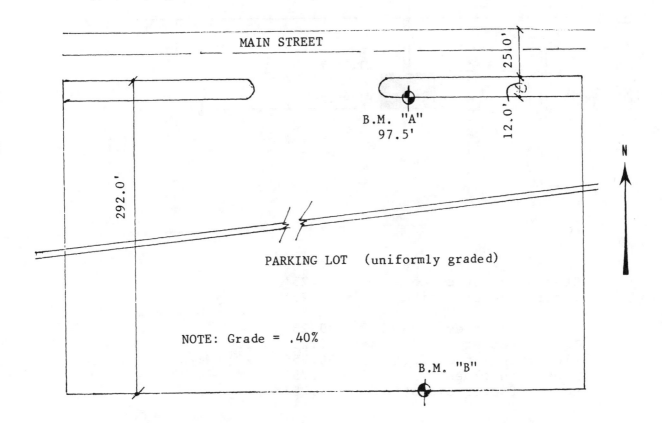

What is the elevation at BM "B" ?

          A.   101.34'
          B.   99.74'
          C.   98.97'
          D.   98.62'

ANSWER SHEET

TRANSIT PROBLEM QUIZ NO. 1

1.   ANSWER D

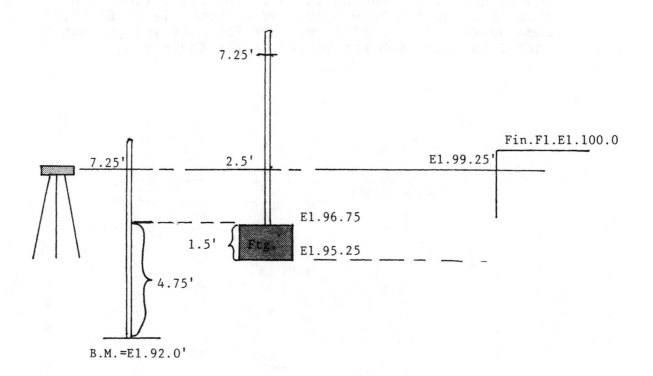

Bench mark elevation   92.00'
Stadia rod at bench mark   7.25'
Instrument elevation   = 99.25'

Instrument elevation   99.25'
Top of footer   2.25'
Elevation top of footer   = 96.75'
Footer thickness   1.75'
Bottom footer elevation   = 95.25'

Finish floor elevation   100.00'
Bottom footer elevation   95.25'
Answer   = 4.75'

ANSWER SHEET

TRANSIT PROBLEM QUIZ NO. 1

2. ANSWER C

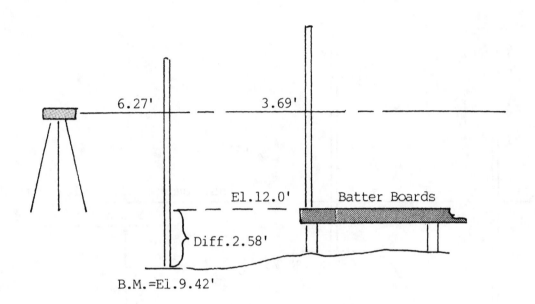

Bench mark elevation   9.42'
Stadia rod   6.27'
Instrument elevation = 15.69'

Instrument elevation   15.69'
Batterboard elevation   12.00
Answer   3.69'

ANSWER SHEET

TRANSIT PROBLEM QUIZ NO. 1

3.   ANSWER C

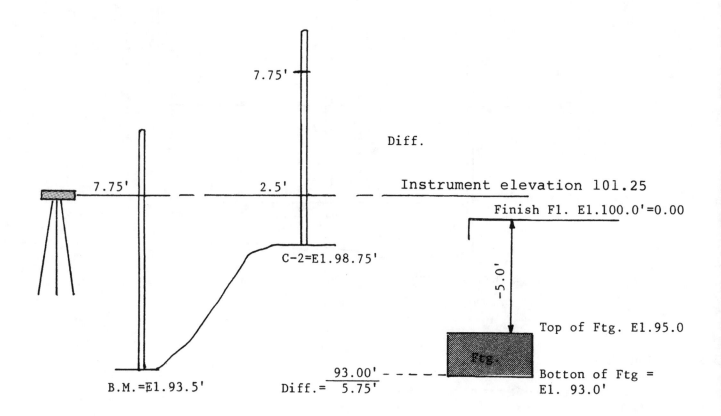

Benchmark elevation		93.50'
Stadia rod at benchmark		7.75'
Instrument elevation	=	101.25'
Instrument elevation		101.25'
Column stadia rod		2.50'
Column elevation	=	98.75'
Finish floor		100.00'
Top of footer		5.00'
Bottom of footer		2.00'
Bottom of footer elevation	=	93.00'
Column elevation		98.75'
Bottom of footer elevation		93.00'
Answer		5.75'

ANSWER SHEET

TRANSIT PROBLEM QUIZ NO. 2

1. ANSWER B

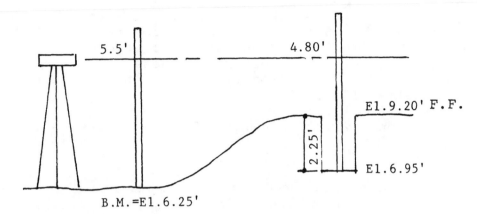

B.M.=El.6.25'

Benchmark elevation	6.25'	Batter board elevation	9.2'
Stadia rod at benchmark	5.50'	Bottom of footer	2.25'
Instrument elevation =	11.75'	Footer elevation =	6.95'

Instrument elevation	11.75'
Footer elevation	6.95'
Answer	4.80'

2. ANSWER C

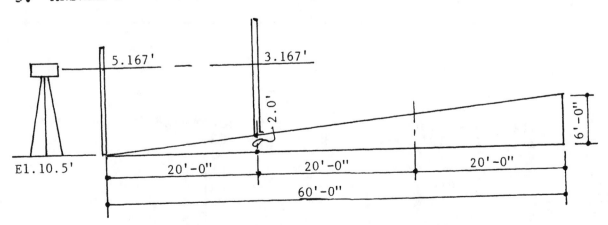

Finish Floor.

3. ANSWER D

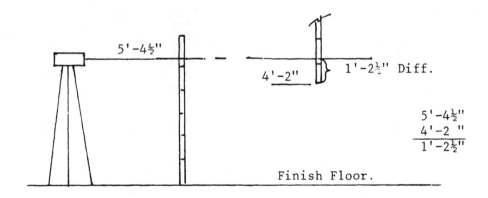

ANSWER SHEET

TRANSIT PROBLEM QUIZ NO. 3

1. ANSWER 1.12' Difference

Benchmark	127.52'	Finish floor elevation	131.36'
Stadia rod at bench.	8.55'	Less top of footer	8.00"
Instrument elev.	136.07'	Elev. at top of footer	123.36'
		Footer thickness	2.67'
		Elev. bottom of footer	120.69

Instrument elev. 136.07'
Stadia at mark 2   14.25'
Mark 2 elevation   121.82'-elev.at bottom of footer = 1.12'

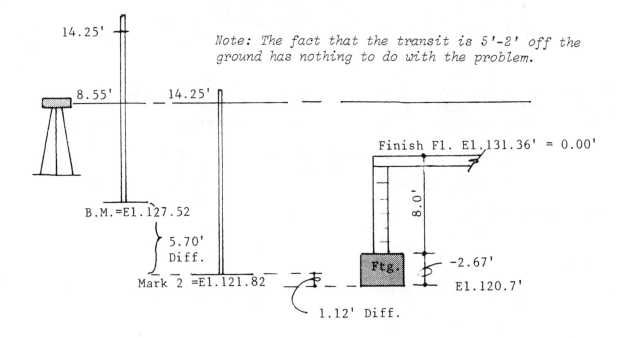

Note: The fact that the transit is 5'-2' off the ground has nothing to do with the problem.

2. ANSWER D

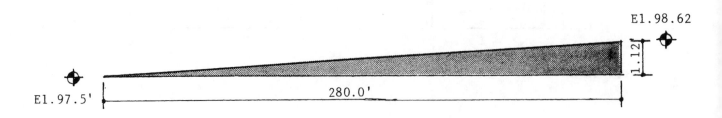

The grade rises .40% per foot therefore 280' x .40% = a rise of 1.12'. EL at BM "B" = 98.62'

SLOPED WOOD ROOFS

When estimating the quantities of lumber for roofs, the
terminology of the component parts of the roof must first be
understood. (See FIG. 1.)

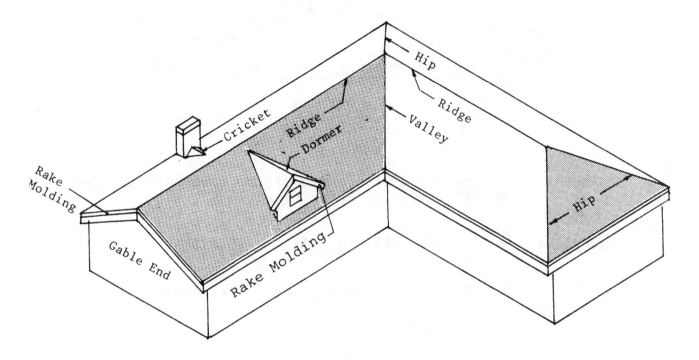

FIG. 1

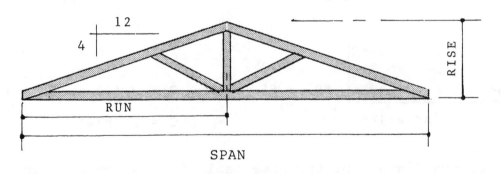

FIG. 2

FIG. 2 shows a cross section of a typical sloped roof.
WALKER'S ESTIMATOR'S REFERENCE BOOK, p 11.7, Table of "Lengths
of Common, Hip, and Valley Rafters per 12 Inches of Run"
refers to "Pitch of Roof", column 1, which is the ratio of
the span to the rise, and "Rise and Run or Cut", column 2,
which is the ratio of the mid-point of the span to the rise.
Architectural plans will symbolize the rise to run as (⊤──)
which indicates a ratio of rise to run of 4 to 12.

Column 3 of the same table is the "Length in Inches Common Rafter per 12 Inches of Run."

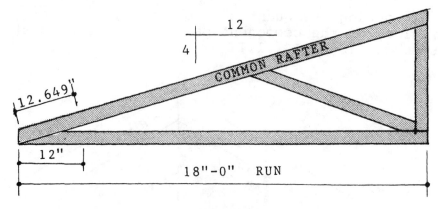

FIG. 3

EXAMPLE: Given: Find the length of the common rafter for an 18'-0" run. (Refer to FIG. 3.)

SOLUTION:

$$\frac{12.649 \times 18}{12} = 18.974'$$

The rise to run is 4 to 12; therefore,

$$\frac{4}{12} : \frac{X}{18} \qquad X = 6$$

Rise is 6   Run is 18

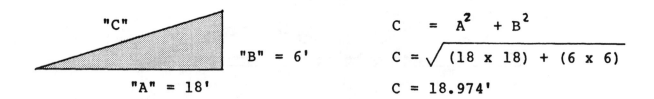

$$C = A^2 + B^2$$

$$C = \sqrt{(18 \times 18) + (6 \times 6)}$$

$$C = 18.974'$$

Column 4 and 5 of the same table is the "Percent Increase in Length of Common Rafter Over Run." Again using a 4 to 12 rise to run, from Column 3 the common rafter was ".649" longer than the run.

$$\frac{.649}{12} = .054\% \text{ increase, or } \frac{12.649}{12} = 1.054\%$$

The 1.054% is used as the multiplier.

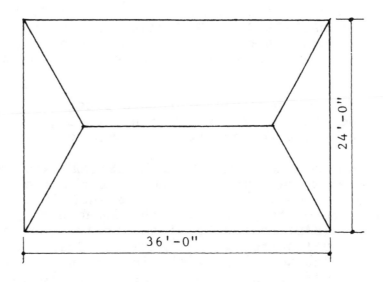

PLAN VIEW

(no scale)

FIG. 4

EXAMPLE: <u>Given</u>: If a 4 to 12 rise to run roof has a flat
           area of 24' x 36' (Plan View), what is the sq ft
           area of the roof? (Refer to Fig.4)

SOLUTION: To find the sq ft of area of the sloped roof,
           36' x 24' = 864 sq ft, which is the flat area
           times the amount of increase. 864' x 1.054 =
           910.7 sq ft of roof area.

Column 6 is the Length in Inches of Hip or Valley Rafter per
12" of Common Rafter.

EXAMPLE: <u>Given</u>: The span is 36'-0". Run to rise is 4 and
           12. What is the length of the valley and/or hip
           rafter?

SOLUTION: The common rafter run is 18'-0". WALKER'S
           ESTIMATOR'S page 11.7 refers to length of run not
           length of common rafters, from column 6, we
           select 17.433 factor.

           $\underline{18 \times 17.433} = 26.15'$ (length of valley and/or hip)
                      12

           (Refer to Figures 1 and 2)

The same table appears in less detail on page 12.3 of WALKER'S ESTIMATOR'S REFERENCE BOOK and the last column is the same as Column 6 on page 11.7, but the factor is in lineal feet of hip or valley per lineal foot of common rafter run.

Often the test writers will refer to pitch as 6/12 or 4/12, which is the same as 1/2 and 1/3; notice that all the even number pitches are reduced to the lowest common denominator.

The table on page 11.7 of WALKER'S ESTIMATOR'S REFERENCE BOOK, is used to determine the area of the roof and is based on the area of plywood with no waste included. When using S4S lumber, refer to page 11.6, third table down "Quantity of Square Edged (S4S) Boards Required Per 100 Square Foot of Surface." Actual board size is shown in Column 2. Column 3 shows the amount added for width in percent. Column 4 is the board measure after the area has been calculated from page 11.7. Notice that Column 4 includes 5% for end cutting and waste. Column 5 is weight per 1000 lin ft of that particular size wood.

EXAMPLE: Given: Using the same roof area as Fig. 4, and using S4S lumber (1 x 8), what is the total amount of lumber required?

SOLUTION: Flat area is 24' x 36' = 864 sq ft. Roof is 4 to 12 rise to run. From page 11.7, column 5, 864 x 1.054 = 910.7 sq ft.

Referring to p 11.6, column 4, (1x8), there are 115 board feet per 100 sq ft of surface; therefore, 115 x 910.7/100 = 1047.3 board feet of 1 x 8, which includes 5% for cutting and waste.

Roof Ridge Lines:

If one roof line intersects another roof line and the span of the two roofs is not the same, the ridge lines will not be in the same plane.

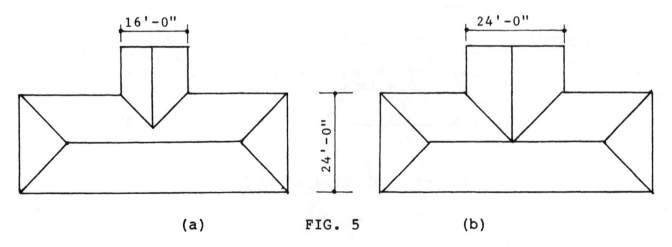

(a)   FIG. 5   (b)

Whereas if two roofs intersect and the span is the same, the ridge lines will meet. (See Fig. 5, (a) and (b)).

Dormers (Roof):

A dormer roof is a small roof with a shorter span than the roof it is projected from and sits entirely on the larger roof. (See Fig. 1.)

The prime purpose of a dormer roof is to give more space to the attic and also more light and ventilation.

Crickets:

The purpose of a cricket is to divert water from a vertical protrusion in a roof such as a chimney. (See Fig. 1.)

Rake Moulding:

Rake moulding is defined in the ARCHITECTURAL AND BUILDING TRADES DICTIONARY, page 359. The type of roof construction in Florida normally incorporates overhang and fascia board construction which would indicate the rake to be the fascia board in the gable end of the roof which would have to be a wider board than the fascia on the eaves side of the roof. (See Fig. 1.)

Board Measure:

In the State Exam board measure is used to determine the volume of wood required for construction. One board foot is lumber with a nominal dimension of 1'-0" x 1'0" x 1" thick.

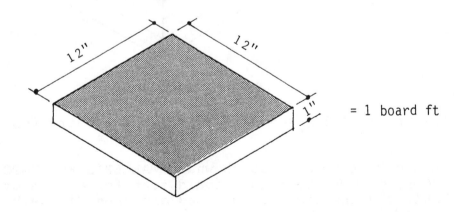

FIG. 6

A 2 x 4 is 2" x 4" or 1" x 8".

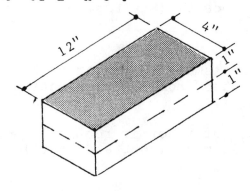

FIG. 7

To figure board feet for a 2 x 4:

$$\frac{2 \times 4}{12} = \frac{8}{12} = .667 \text{ board feet.}$$

A 4 x 12 = 4" x 12" or 1" x 12", 4 times.

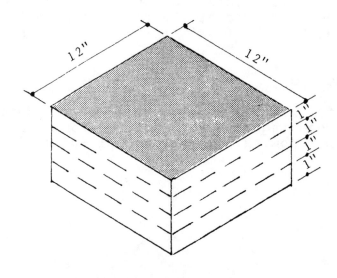

FIG. 8

Example: A 3 x 4 = 3" x 4" = $\frac{12"}{12}$ = 1 board foot

Board Feet Measure is shown in WALKER'S ESTIMATOR'S REFERENCE BOOK, p 11.6. To figure board feet for a size of wood multiply the size of the wood and then divide by 12 to equal board feet.

When determining the quantity of lumber for partitions, formwork, etc., it is always asked for in board feet.

Floors, Joists, Partitions:

When determining the amount of joists or studs in wood
construction it must first be known how they are placed
on center. Most joists and studs are placed 12 inches,
16 inches, or 24 inches on center.

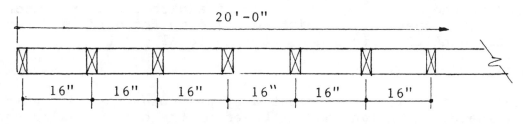

FIG. 9

When the studs are 16 inches on center they are 16 inches
from outer edge to outer edge.

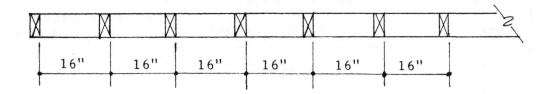

FIG. 10

When counting the number of studs in a given length it
is actually the spaces that are being counted with one
additional stud added for a starter.

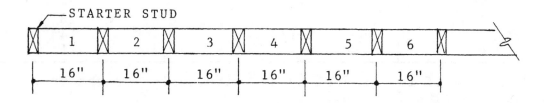

FIG. 11

If there is 20 ft of wall with studs at 16 inches on center,
the number of studs may be determined by dividing the length
by 1.333'; (16"/12" = 1.333').

$$\frac{20}{1.333} = 15 \text{ spaces} + 1 \text{ starter equals } 16 \text{ studs}$$

or the length may be multiplied by .75 (1/1.333' = .75'space
between studs). 20 x .75 = 15 + 1 starter equals 16 studs.

A multiplication factor table may be found in WALKER'S ESTIMATOR'S REFERENCE BOOK, page 11.2.

When estimating wood quantities caution must be taken to see that whenever added pieces are included, such as a double stud at the beginning and/or end of a partition, or a door and window opening, or joists doubled up for girders, that these quantities are added into the total count.

Nails:

The size of nails are indicated by the "penny system" which originated in England and is based on the cost of nails per hundred pounds. The term penny is also denoted by the letter "d" derived from the Greek word Devaries (a coin).

Today nails are still denoted by the penny system "d" but are referred to by length in inches.

Size	Length (in inches)	Approximate No. to the pound
4d	1-1/2	316
5d	1-3/4	271
6d	2	181
7d	2-1/4	161
8d	2-1/2	106
9d	2-3/4	96
10d	3	69
12d	3-1/4	63
16d	3-1/2	49
20d	4	31
30d	4-1/2	24
40d	5	18
50d	5-1/2	14
60d	6	11

Nails are of two general types, wire nails (bright or galvanized) and cut nails.

Types of wire nails are common nails, finish nails, roof nails, shingle nails, double head, gypsum, masonry, ratchet, and T nails.

For more information on nails see:

WALKER'S ESTIMATOR'S REFERENCE BOOK, 22nd ED, 1986
ARCHITECTURAL AND BUILDING TRADES DICTIONARY, 1980
NFPA WOOD CONSTRUCTION DATA, NO. 1, 4 and 6

To find the horizontal projection of
a rafter when the slope length of the
rafter is known.

Slope over 12	Factor
3	0.970
4	0.958
5	0.923
6	0.894
7	0.864
8	0.832
9	0.800
10	0.768
11	0.737
12	0.707
13	0.678
14	0.651
15	0.625
16	0.600
17	0.577
18	0.555
19	0.534
20	0.514

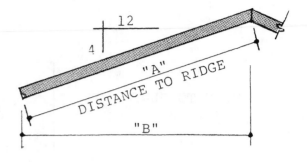

B = A x Factor

A = 15 ft

B = 15 x 0.958

ROOFS - SECOND DAY TYPE QUESTIONS

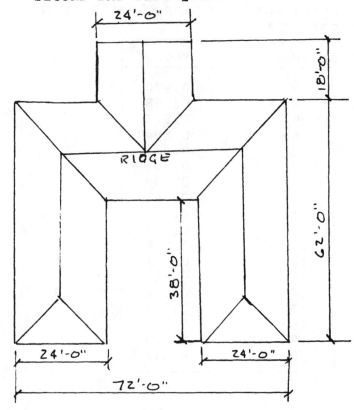

RISE TO RUN
 = 6/12

1. There are a total of _____ hips.

    A. 2   B. 4   C. 5   D. 6   E. 8

2. There is a total of _____ lin ft of hip.

    A. 36   B. 48   C. 72   D. 108   E. 124

3. There is a total of _____ valleys.

    A. 2   B. 4   C. 5   D. 6   E. 8

4. There is a total of _____ lin ft of valley.

    A. 36   B. 48   C. 72   D. 108   E. 124

5. The roof is a total of _____ in area.

    A. 35.5 squares   B. 29.8 sqs   C. 32 sqs   D. 44.45 sqs

6. A total of _____ rake molding is required.

    A. 13.4'   B. 24'   C. 26.8'   D. 42.5'   E. 53.7

7. The total length of ridge is _____ .

    A. 130   B. 160   C. 127   D. 138   E. 154

ANSWER SHEET

ROOFS - SECOND DAY TYPE QUESTIONS

RISE TO RUN
= 6/12

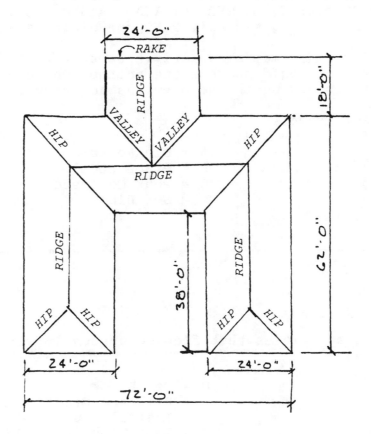

REFER TO PAGE 12.33 OR 11.7 OF WALKER'S ESTIMATOR'S.

1. (D)   There are six hips.

2. (D)   6 hips x 12' x 1.5 or 18 = 108 lin ft of hip.

3. (B)   There are four valleys.

4. (C)   4 valleys x 12' x 1.5 or 18 = 72 lin ft of valley.

5. (D)   (72' + 38' + 38' + 18") x 24' = 3984 sq ft
         3984 x 1.118 / 100 = 44.54 squares

WALKER'S ESTIMATOR'S PAGE 728 – one square contains 100 sq ft

6. (C)   24' X 1.118 = 26.832'

7. (E)   18' + 12' + (72' - 12' - 12') + (38' x 2) = 154 ft

THE CRITICAL PATH METHOD

The Critical Path Method (CPM) is an organized planning
system of scheduling work in a sequence graphically so as to
plot the completion of a particular work project whether it
be large or small. The Critical Path Method (CPM) is also
known as a "Flow Chart". See WALKER'S ESTIMATOR'S REFERENCE
BOOK, Chapter 5, or THE CRITICAL PATH METHOD, a self-study
text, by Jose D. Mitrani, P.E. for diagrams and discussion.

The Flow Chart may show activities that will flow concurrently, or that would precede one another where one activity
cannot proceed until the previous activity is completed.

    For Example:

            A = Dig Footers
            B = Place Steel
            C = Pour Concrete
            D = Set Block

FIG. 1

Whereas activities that flow concurrently will diagram as
follows:

            D = Set Block
            E = Fill and Grade
            F = Install Plumbing
            G = Test Plumbing
            H = Cover Plumbing
            I = Install Electrical
            J = Cover Electrical
            K = Grade Outside Perimeter
            L = Pour Concrete

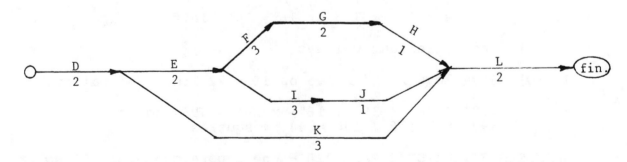

FIG. 2

In Fig. 2 the Plumbing(F-G-H), Electrical(I-J), and Grading (K) may all flow concurrently. The Critical Path is the series of tasks which will take the longest. In Fig. 2 the plumbing is the Critical Path, although if the electrical required an additional three days the Critical Path would change. In Fig. 2 the plumber requires six days and the electrician four days, therefore the electrical has two days of flow time, or he does not have to start until two days after the plumber starts.

It is recommended that Chapter 5 of WALKER'S ESTIMATOR'S REFERENCE BOOK be fully understood and that all of the problems be worked out completely in the self-study text, THE CRITICAL PATH METHOD, Jose D. Metrani, P.E., as the Critical Path Method is an important part of the State Exam and speed and accuracy are required to do Critical Path problems properly.

RELATIVE HUMIDITY

Problems related to condensation in structures require a knowledge of terms. Realtive humidity is the term that denotes the amount of moisture in dry air at a given dry-bulb temperature relative to what the air could hold at that temperature if it were totally saturated with moisture; it is an expression of percentage of saturation. Its symbol is RH or %RH.

The point at which water vapor in the atmosphere begins to condense is termed the dew-point temperature of the air. When a glass of ice water begins to sweat in a warm, moist room or when a film of moisture forms on a cold pipe in a warm space, the "local" air reaches its dew-point temperature and begins to condense. In an air conditioning system, the air passing through a cooling coil will reach a condition below its dew-point and moisture, in the form sweat on the face of the cold coil, will appear. That is how room air is dehumidified; moisture from the air collects on the coils, drips down on a collector pan and drains down the condensate pipe.

The most common instrument for measuring relative humidity (RH) is the sling psychrometer. The sling psychrometer consists of two speparate thermometers attached to a thin metal plate or frame that can be whirled in the air. The bulb of one thermometer is covered by a wet wick, or sock; this is called the wet-bulb. Whirling the psychrometer in the air causes evaporation to occur at the wet wick thus lowereing the wet-bulb temperature --cooling by evaporation. The two thermometers will now read differently. The difference in temperature between the two thermometers determines the relative humidity of the air, see the CONCRETE MASONARY HANDBOOK, page 56. Also, see the Relative Humidity Table below.

Insulation, with vapor barrier, is used to prevent the formation of condensation in buildings. Together with proper ventilation, insulation is a requirement for any properly constructed structure for condesation protection and energy conservation.

RELATIVE HUMIDITY TABLE

DIFFERENCE BETWEEN THE DRY AND WET BULB TEMPERATURES, °F
(WB Depression)

% RELATIVE HUMIDITY

Dry Bulb	0	1	2	3	4	5	6	7	8	9	10	11	12	13	14	15	16	17	18	19	20	21	22	23	24	25	26	27	28	29	30	31	32	33	34	35	36
30	100	89	78	67	57	47	36	26	17	7																											
35	100	91	82	73	65	54	45	37	28	19																											
40	100	92	84	76	68	60	53	45	38	29	22																										
45	100	92	85	78	71	64	58	51	44	38	32	25	19																								
50	100	93	87	80	74	67	61	55	50	44	38	33	27	22	16																						
55	100	94	88	82	76	70	65	59	54	49	44	39	34	29	24	19																					
60	100	94	89	83	78	73	68	63	58	53	48	44	39	35	30	26	22	18	14																		
65	100	95	90	85	80	75	70	66	61	56	52	48	44	39	35	31	28	24	20	17	13	10															
70	100	95	90	86	81	77	72	68	64	60	55	52	48	44	40	36	33	29	26	23	19	16	13	10	7	4	1										
75	100	96	91	87	82	78	74	70	66	62	58	55	51	47	44	40	37	34	30	27	24	21	18	16	13	10	7	5	2								
80	100	96	92	88	83	79	75	72	68	64	61	57	54	50	47	44	41	38	35	32	29	26	23	20	18	15	12	10	8	6	3	1					
85	100	96	92	88	84	80	77	73	69	66	62	59	56	52	49	47	44	41	38	36	33	30	28	25	23	20	18	15	13	11	9	6	4	2			
90	100	96	92	89	85	81	78	74	71	68	65	61	58	55	52	49	47	44	41	39	36	34	32	29	26	24	22	20	17	15	13	11	9	7	5	3	2
95	100	96	93	90	86	82	78	75	72	69	66	63	60	58	55	51	49	47	44	42	39	37	35	32	30	28	26	24	22	19	17	15	14	12	10	8	6
100	100	97	93	90	87	83	80	77	74	71	68	65	63	60	57	55	53	50	48	46	44	42	40	38	36	34	31	30	28	26	24	22	20	19	17	15	14
105	100	97	93	90	87	84	81	78	75	72	69	66	64	61	58	56	53	51	49	47	45	42	40	38	36	34	31	30	28	26	24	22	20	19	17	15	14
110	100	97	94	91	88	85	82	79	76	73	70	67	65	62	60	57	55	53	50	48	46	44	42	40	38	36	34	32	30	28	26	24	23	21	20	17	17
115	100	97	94	91	88	86	83	80	77	75	72	69	66	64	62	59	57	55	53	51	49	47	45	43	41	39	37	35	33	31	30	28	26	24	23	21	20
120	100	97	94	91	88	86	83	81	78	76	73	70	67	65	63	60	58	56	54	52	50	49	47	45	42	41	40	38	36	33	32	30	28	27	25	24	22
125	100	97	94	91	89	86	84	82	79	77	74	72	69	66	64	62	59	57	56	53	51	49	47	45	43	42	40	38	37	35	33	31	29	28	26	26	24
130	100	97	94	92	89	86	84	82	80	78	75	73	70	68	65	63	62	60	58	56	54	51	50	48	46	44	42	40	38	36	34	33	31	29	28	28	27
135	100	97	94	92	89	86	84	81	79	77	74	72	70	67	65	63	61	59	57	55	53	52	50	48	45	43	41	40	38	36	35	33	31	30	30	30	28
140	100	97	95	92	89	87	84	82	79	77	75	73	71	68	66	64	62	60	58	56	55	53	51	49	48	46	44	42	40	38	36	34	33	31	31	31	30

DIFFERENCE BETWEEN THE DRY AND WET BULB TEMPERATURES, °F

AIR DRY BULB TEMPERATURE, °F

EXAMPLE: Assume a dry bulb temperature of 90F and a wet bulb temperature of 80F. The difference is 10 degrees. Enter the chart at 90F in the left quadrant and move out to the 10 degree difference column. At that intersection the relative humidity (RH) is shown to be 65%. This table was constructed from data compiled by the U.S. Weather Bureau.

HEAT TRANSMISSION THROUGH BUILDINGS

1. Heat transmission is composed of Conductance and
    Resistance. The conductance is the amount of heat
    or cold that will flow through a surface, and the
    resistance is the material that will slow down the
    flow of heat or cold.

2. The amount of flow is calculated as Btuh. Btu is
    defined as British Thermal Unit which is the amount
    of heat to raise one pound of water one degree F),
    and is calculated for one hour (H) which is Btuh.

    > Btuh = Area of Surface x Total U x Temperature
    > difference between inside and outside
    > surfaces.
    > (surface area is in square feet)

3. Total U is derived by computing all of the resistances
    (R) of a total surface that is being penetrated, then
    taking the reciprocal of the total summation of the
    (Accumulative) R's. R is derived by laboratory tests
    wherein calculations are made to determine how much
    conductance (K or C) will penetrate the surface. Each
    particular material of a total surface has its own K or C
    value. K being the amount of Btu per sq ft per hour per
    degree F that will penetrate one inch of thickness from
    surface per degree F that will penetrate from surface to
    surface regardless of thickness. The R value is derived
    by the reciprocal (1/K or 1/C) of the conductance.

FIGURE 1

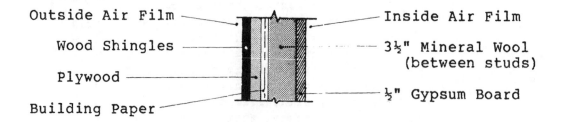

Outside Air Film — Inside Air Film

Wood Shingles — 3½" Mineral Wool
(between studs)

Plywood

Building Paper — ½" Gypsum Board

The C and R values for Fig. 1 are as follows:

Wood Shingles	C = 1.064	R = 1/1.064 = .94
Plywood	C = 1.075	R = 1/1.075 = .93
Building Paper	C = 8.333	R = 1.8.333 = .12
3-1/2" Mineral Board	C = .091	R = 1/.091 = 11.00
1/2" Gypsum Board	C = 2.273	R = 1/2.273 = .44

<u>Note</u>: The lower the R the higher the C or K.

4. To calculate the total U of the surface in Fig. 1 add all of the R values and divide into one:

$$U = \frac{1}{R_1 + R_2 + R_3 + R_4 + R_5}$$

In order to finalize total U or total R (Cumulative R), inside and outside air film must be included. You cannot use U in the formula Btuh = A x U x TD without including the R for inside and outside air film in the above formula. Also remember that R's are the only factors that can be added.

Inside Air Film: R = .68

Outside Air Film: R = .17

The above factors must always be included in calculation of Total (Cumulative) R and/or U.

5. Applying the above formula to Figure 1:

Total R = .17 + .94 + .93 + .12 + 11.0 + .44 + .68 = 14.28

Total U = $\frac{1}{14.28}$ = .07

6. Assuming 1000 sq ft of wall, with an outside temperature of 95°F, and an inside temperature of 78°F:

Btuh = Area x U x Temp. Diff.

Btuh = 1000 sq ft x .07 x 17 = 1190

The amount of heat that would flow through 1000 sq ft of wall at a 17°F Temperature difference.

7. For heat flow through glass, refer to the CONCRETE MASONRY HANDBOOK, page 54, (1.13 Btu per sq ft per F degree per hour) is the U Value for standard glass. Standard glass is single pane, winter condition. In the State exam 1.13 is considered to be the Total (Cumulative R) which is thermal glass. The reciprocal of 1.13 equals .884 or the U Value (1/R = U, 1/1.13 = .884). This U Value, .884, must then be used in the formula Btuh = A x U x TD. If an R value is not referred to as Total or Cumulative R, then the two air films must be added to the R which then makes it total (Cumulative) R. —— U can be derived by dividing the Total or Cumulative R into one.

1/Total or Cumulative R = U

UNDERSTANDING INSULATION AS A HEAT TRANSFER RESISTANT

Whether insulation is used for cold storage walls, chilled water lines, steam lines, residential attics, or curtain walls, it is based on the principle of slowing down the rate of flow of heat through a substance. This heat transmission depends on the material and thickness of insulation thus constituting a condition of thermal conductivity or transfer coefficient.

GLOSSARY OF TERMS

k factor: The amount of heat in Btu's transmitted through 1 sq ft of 1 in. thick material for a difference of 1°F per hour from surface to surface. See Figure 1.

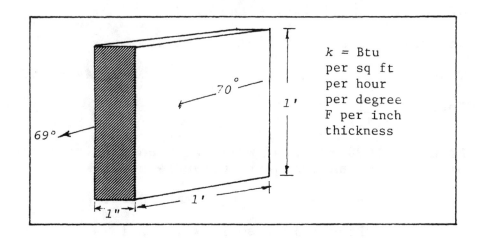

FIGURE 1

THERMAL CONDUCTIVITY, K

The K factor for mineral wool blankets is 0.27; therefore, the transferrence of heat through a 1 in thick mineral wool blanket would be 0.27 Btu/hr per sq ft per degree difference between the two surfaces.

C factor: Thermal conductance, the amount of heat in
Btu's that will pass through 1 sq ft per 1
hr between the two surfaces of any material,
or combination of materials of construction
being considered, for a difference of 1°F;
not per inch of thickness. The heating, air
conditioning, and refrigerating engineer
deals mainly with compound walls of more than
one material, and makes considerable use of
conductance factor. Figure 2 illustrates the
C factor.

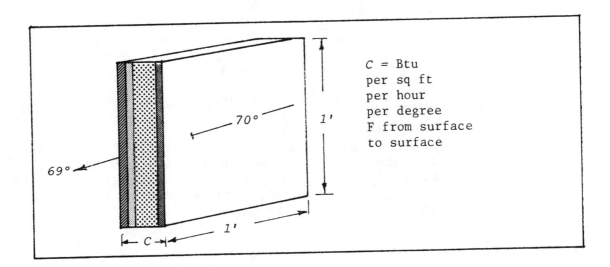

FIGURE 2

THERMAL CONDUCTIVITY, C

U factor: Sometimes called the U value, designates
the total or overall transmission of heat
in Btu's in 1 hr per sq ft of area for a
difference in temperature of 1 degree F
between the air on one side to the air on
the other side of any construction. The
term is applied to the usual combinations
of materials for walls, roofs, floors or
ceiling, as well as for single materials
such as glass windows. Figure 3 illustrates the U factor.

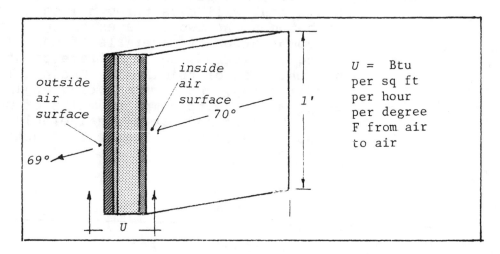

FIGURE 3

OVERALL COEFFICIENT, U

R: Thermal resistance, the reciprocal of K, C, or U
expressed as degree F per Btu per 1 hr per sq ft.
The reciprocal of any number is 1 divided by the
number. Thus, a wall with a U value of 0.37 will
have a resistance of 2.7.

$$\frac{1}{0.37} = 2.7 \ R$$

Industrial insulation: Includes the application of
thermal insulation to cold storage rooms, pipes,
tanks and vessels, and ducts as opposed to general
building insulation.

Water vapor barrier: A material that will reduce the
rate of volume of water vapor under specific
conditions. Water vapor in air is a gas, which can
move through a material under differences in its own
vapor pressure independently of the air with which
it is mixed.

Surface condensation: The condensation that may appear on the warm side of insulated pipes, ducts, rooms, etc., when the water vapor in the air comes in contact with a surface whose temperature is lower than the dewpoint of the air. A sufficient thickness of insulation whould be used to keep the temperature of the surface of the insulation higher than the dewpoint temperature.

Calculating Overall Coefficients

To calculate the U value of a floor, wall, ceiling or roof it is necessary to know the K or C of all of the component material making up the section. The conductivities and/or conductances are not additive, but must be converted to resistances. Values for various materials may be found in WALKER'S INSULATION TECHNIQUES AND ESTIMATING HANDBOOK pp 125-128. Once the values for K or C have been determined and the resistances are calculated, the U value may be found by the formula:

$$U = \frac{1}{R + R + R + R + R + R}$$

EXAMPLE: Find the U value of a frame wall consisting of 1/2 in. gypsum board, 3-1/3 in. mineral wool blanket, 3/4 in. plywood, building paper, and wood shingles.

SOLUTION:

Section	Resistance R	
Air film (outside, 15 mph)	1/6.0 =	0.17
Gypsum board, 1/2 in	1/2.25 =	0.44
Mineral Wool, 3-1/2 in	1/0.09 =	11.00
Plywood, 3/4 in	1/1.07 =	0.93
Building paper	1/8.35 =	0.12
Wood shingles	1/1.06 =	0.94
Air film (inside still air)	1/1.46 =	0.68

Total resistance of each sq ft = 14.28

$$\text{Therefore, } U = \frac{1}{R + R + R + R + R + R + R}$$

$$= \frac{1}{14.28} = 0.07 \text{ Btu per hr per sq ft per deg F}$$

Now, if the wall used in the above example was 30 ft x 10 ft and the temperature difference between the inside and outside was 20°F, then:

300 sq ft x 20°F x .07 = 420 Btuh heat transmission. The exammple is illustrated graphically in Figure 4.

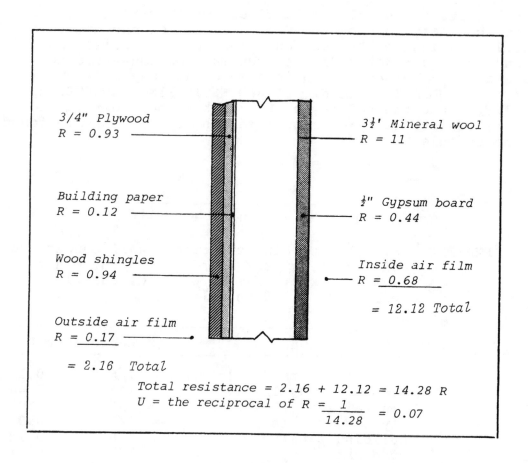

3/4" Plywood
R = 0.93

Building paper
R = 0.12

Wood shingles
R = 0.94

Outside air film
R = 0.17

= 2.16 Total

3½' Mineral wool
R = 11

½" Gypsum board
R = 0.44

Inside air film
R = 0.68

= 12.12 Total

Total resistance = 2.16 + 12.12 = 14.28 R
U = the reciprocal of R = $\frac{1}{14.28}$ = 0.07

FIGURE 4

CALCULATING COEFFICIENT U

HEAT TRANSMISSION FACTORS FOR BUILDING MATERIALS

MATERIAL	DESCRIPTION	CONDUC-TIVITY K#	CONDUCT-ANCE c +
BUILDING BOARDS	ASBESTOS-CEMENT BOARD....................................	4.0	
	GYPSUM OR PLASTER BOARD...1/2 IN.		2.25
	PLYWOOD...	0.80	
	PLYWOOD...3/4 IN.		1.07
	SHEATHING (IMPREGNATED OR COATED).............	0.38	
	SHEATHING (IMPREGNATED OR COATED) 25/32 IN.		0.49
	WOOD FIBER—HARDBOARD TYPE........................	1.40	
INSULATING MATERIALS	BLANKET AND BATT:		
	MINERAL WOOL FIBERS (ROCK, SLAG, OR GLASS)..	0.27	
	WOOD FIBER...	0.25	
	BOARDS AND SLABS:		
	CELLULAR GLASS......................................	0.39	
	CORKBOARD..	0.27	
	GLASS FIBER...	0.25	
	INSULATING ROOF DECK...2 IN.		0.18
MASONRY MATERIALS	LOOSE FILL:		
	MINERAL WOOL (GLASS, SLAG, OR ROCK).........	0.27	
	VERMICULITE (EXPANDED).............................	0.46	
	CONCRETE:		
	CEMENT MORTAR.......................................	5.0	
	LIGHTWEIGHT AGGREGATES, EXPANDED SHALE, CLAY, SLATE, SLAGS; CINDER; PUMICE; PERLITE; VERMICULITE..............................	1.7	
	SAND AND GRAVEL OR STONE AGGREGATE.........	12.0	
	STUCCO..	5.0	
	BRICK, TILE, BLOCK, AND STONE:		
	BRICK, COMMON..	5.0	
	BRICK, FACE..	9.0	
	TILE, HOLLOW CLAY, 1 CELL DEEP, 4 IN.........		0.90
	TILE, HOLLOW CLAY, 2 CELLS, 8 IN.		0.54
	BLOCK, CONCRETE, 3 OVAL CORE:		
	SAND & GRAVEL AGGREGATE...4 IN.		1.40
	SAND & GRAVEL AGGREGATE...8 IN.		0.90
	CINDER AGGREGATE............ 4 IN.		0.90
	CINDER AGGREGATE............. 8 IN.		0.58
	STONE, LIME OR SAND...............................	12.50	
PLASTERING MATERIALS	CEMENT PLASTER, SAND AGGREGATE..................	5.0	
	GYPSUM PLASTER:		
	LIGHTWEIGHT AGGREGATE...1/2 IN..................		3.12
	LT. WT. AGG. ON METAL LATH...3/4 IN............		2.13
	PERLITE AGGREGATE..................................	1.5	
	SAND AGGREGATE.....................................	5.6	
	SAND AGGREGATE ON METAL LATH 3/4 IN........		7.70
	VERMICULITE AGGREGATE	1.7	
ROOFING	ASPHALT ROLL ROOFING		6.50
	BUILT-UP ROOFING...3/8 IN............................		3.00
SIDING MATERIALS	ASBESTOS-CEMENT, 1/4 IN. LAPPED....................		4.76
	ASPHALT INSULATING (1/2 IN. BOARD)		0.69
	WOOD, BEVEL, 1/2 X 8, LAPPED		1.23
WOODS	MAPLE, OAK, AND SIMILAR HARDWOODS	1.10	
	FIR, PINE, AND SIMILAR SOFTWOODS	0.80	
	FIR, PINE & SIM. SOFTWOODS 25/32 IN.............		1.02

#Conductivity given in Btu in. per hr sq ft F
+Conductance given in Btu per hr sq ft F

INSULATION MATERIALS CHART

TYPE	COMMENTS	APPLICATION
BATTS/BLANKETS Preformed glass fiber or rock wool with or without vapor barrier backing.	Fire resistant, moisture resistant, easy to handle for do-it-yourself installation, least expensive and most commonly available.	Unfinished attic floor, rafters, underside of floors, between studs.
FOAMED IN PLACE Plastic installed as a foam under pressure. Hardens to form insulation. 1.) Urethane 2.) Ureaformaldehyde 3.) Polystyrene	1.) Has highest R-value. If ignited, burns explosively and emits toxic fumes. Should be covered with ½" gypsum wallboard to assure fire safety. 2.) Fire resistant, high R-value, first choice of many experts. Requires installation by reliable, experienced contractor (as all foams do). 3.) Lacks fire resistance as does urethane and has lower R-value than urethane.	Finished frame walls; floors; ceilings.
RIGID BOARD 1.) Extruded polystyrene bead 2.) Extruded polystyrene 3.) Urethane 4.) Glass fiber	All have high R-values for relatively small thickness. 1.2.3.) Are not fire resistant, require installation by contractor with ½" gypsum board to insure fire safety. 3.) Is its own vapor barrier; however, when in contact with liquid water, it should have a skin to prevent degrading. 1.,4.) Require addition of vapor barrier. 2.) Is its own barrier.	Basement walls; new construction frame walls; commonly used as an outer sheathing between siding and studs.
LOOSE FILL (POURED-IN) 1.) Glass fiber 2.) Rock wool 3.) Treated cellulosic fiber	All easy to install; require vapor barrier bought and applied separately. Vapor barrier may be impossible to install in existing walls. 1.,2.) Fire resistant, moisture resistant. 3.) Check label to make sure material meets federal specifications for fire and moisture resistance and R-value.	Unfinished attic floor; uninsulated existing walls.
LOOSE FILL (BLOWN-IN) 1.) Glass fiber 2.) Rock wool 3.) Treated cellulosic fiber	All require vapor barrier bought separately; all require space to be filled completely. Vapor barrier may be impossible to install in existing walls. 1.,2.) Fire resistant, moisture resistant. 3.) Fills up spaces most consistently. When blown into closed spaces, has slightly higher R-value; check label for fire and moisture resistance and R-value.	Unfinished attic floor; finished attic floor; finished frame walls; underside of floors.
CAULKING COMPOUNDS 1.) Oil or resin base 2.) Latex, butyl, polyvinyl base 3.) Elastomeric base: Silicones, polysulfides, polyurethanes	1.) Lowest cost, least durable—replacement time approximately 2 years. 2.) Medium priced, more durable—look for guarantees on time of durability. 3.) Most expensive, most durable. Urethanes at $2-$3/cartridge. Recommended by some experts as a best buy. Note: Lead base caulk is not recommended because it is toxic.	At stationary joints; exterior window and door frames; whenever different materials or parts of building meet.
WEATHERSTRIPPING 1.) Felt or foam strip 2.) Rolled vinyl-with or without metal backing 3.) Thin spring metal 4.) Interlocking metal channels s	1.) Inexpensive, easy to install, not very durable. 2.) Medium priced, easy to install, durable, visible when installed. 3.) More expensive, somewhat difficult to install, very durable, invisible when installed. 4.) Most expensive, difficult to install, durable, excellent weather seal.	At moving joints; perimeter of exterior doors; inside of window sashes.

R—VALUES

	BATTS or BLANKETS		LOOSE FILL (POURED-IN)			RIGID PLASTIC FOAMS			
	Glass Fiber	Rock Wool	Glass Fiber	Rock Wool	Cellulosic Fiber	Urethane	U-F	Styrene	
R-11	3½"-4"	3"	5"	4"	3"	1½"	2"	2¼"	R-11
R-13	4"	4½"	6"	4½"	3½"	2"	2½"	2¾"	R-13
R-19	6"-6½"	5¼"	8"-9"	6"-7"	5"	2¾"	3¾"	4¼"	R-19
R-22	6½"	6"	10"	7"-8"	6"	3"	4"	4½"	R-22
R-26	8"	8½"	12"	9"	7"-7½"	3¾"	5"	5½"	R-26
R-30	9½"-10½"	9"	12"-13"	10"-11"	8"	4½"	5¾"	6½"	R-30
R-33	11"	10"	15"	11"-12"	9"	4¾"	6½"	7¼"	R-33
R-38	12"-13"	10½"	17"-18"	13"-14"	10"-11"	5½"	7½"	8½"	R-38

165

HEAT-LOAD PROBLEM NO. 1

1. From Plan No. 3 (Plan Book for Fl. Contractors)
determine the total load on all of the North Glass.

DESIGN CONDITIONS:

Indoor   72°F
Outdoor ambient   40°F
Outdoor wind velocity   15 MPH
Chill factor   10°F
Cumulative R of glass   1.13

Do not deduct for aluminum store front.

Select the closest answer:

    A. less than 48,000 Btuh
    B. between 48,000 and 50,000 Btuh
    C. between 50,000 and 52,000 Btuh
    D. between 52,000 and 54,000 Btuh
    E. greater than 54,000 Btuh

2. From Plan No. 3 (Plan Book for Fl. Contractors) determine
the total load of all concrete block in third floor South
Wall, (conditioned space).

DESIGN CONDITIONS:

Indoor   78°F DB, 50% RH
Outdoor ambient   95°F DB, 79°F WB

Assume a Wall Construction of the following material:

    8" concrete block
    gypsum board
    1 inch insulation in furring strips

Select the closest answer:

    A. less than 2750 Btuh
    B. between 2750 and 2850 Btuh
    C. between 2850 and 2950 Btuh
    D. between 2950 and 3050 Btuh
    E. greater than 3050 Btuh

HEAT-LOAD PROBLEM NO. 2

1. GIVEN: All of the North glass of an office building is equal to 1200 sq ft. Calculate the amount of heat loss through the glass. Cumulative R of the glass 1.13.

   DESIGN CONDITIONS

Indoor Design	70°F
Outdoor Ambient	42°F
Chill Factor	20°F

   A. 27,562
   B. 28,665
   C. 29,568
   D. 32,968
   E. 37,968

2. GIVEN: An area of decorative brick installed in place of window glass in a store front. The area of brick is 10'-0" x 40'-0". The C value of common brick is 1.25. What is the loss through the wall?

   DESIGN CONDITIONS

Indoor Design	76°F
Outdoor Ambient	95°F
Wind Velocity	15 MPH

   A. 3952
   B. 4636
   C. 5560
   D. 9500

ANSWER SHEET

HEAT-LOAD PROBLEM NO. 1

1. From Sheet 2 of plans, the total length of N glass =
   26' x 5 = 130'. Five columns at 20" = 20" x 5 / 12" =
   8.33'. 130' - 8.33' = 121.67'.
   From Sheet 6, all glass is 11' high. 121.67' x 11' =
   1338.37 sq ft (first floor).

   From Sheet 3, second and third floors have 20 windows
   and from Sheet 5, Note: Masonry opening for windows
   is 3' x 10.58' = 634.8 sq ft.

   Total area of glass = 1338.37 + 634.8 = 1973.17 sq ft.

   Cumulative R = 1.13; therefore, U = 1 / 1.13 = .88

   TD = 72°F - 40°F = 32°F

   Btuh = Area x U x TD = 1973.17 x .88 x 32 = 55,564.46

   ANSWER E

2. From Sheet 3 of plans S wall = 26' x 5 spaces - 5
   columns at 20" = 100 / 12 = 8.33' = 130 - 8.33 =
   121.67' + 16' x 2 = 1.67 for columns = 152 sq ft of
   block.

   From Sheet 10, wall is 9.0' high to conditioned ceiling
   9.0' x 152 = 1368 sq ft of concrete block wall, less 10
   windows at 27.0 sq ft = 270.0 sq ft = 1098.0 sq ft of
   block.

   From page 52 of the CONCRETE MASONRY HANDBOOK, wall is
   equal to D1 = U of .148.

   Btuh = Area x U x TD = 1098.0 x .148 x 17 = 2762.5

   ANSWER B

ANSWER SHEET

HEAT-LOAD PROBLEM NO. 2

1. Cumulative R = 1.13; therefore,

   $$U = \frac{1}{1.13} = 0.885$$

   TD = 70 - 42 = 28

   Btuh = A x U x TD

   Btuh = 1200 x .885 x 28 = 29,736

   ANSWER C

2. To the C value of brick must be added the inside air film of 0.68 and outside air film of 0.17.

   $$R = .68 + \frac{1}{1.25} + .17 = 1.65$$

   $$U = \frac{1}{1.65} = .61$$

   TD = 95 - 76 = 19

   Btuh = A x U x TD

   Btuh = 400 x .61 x 19 = 4636

   ANSWER B

CONCRETE, CONCRETE REINFORCEMENT, AND MASONRY

A good portion of the State Contractors exam is devoted to concrete and its components.

The three required reference books pertaining to concrete are:

> DESIGN AND CONTROL OF CONCRETE MIXTURES, Portland Cement Association.
> CONCRETE MASONRY HANDBOOK, F.A. Randall & W.C. Panarese, Portland Cement Association.
> PLACING REINFORCING BARS, Concrete Reinforcing Steel Institute.

These books are briefly described in this section. They are very important books to have complete knowledge of, as there are many questions asked regarding concrete placement, reinforced concrete, and concrete masonry.

When calculating concrete volume be sure not to include the same area more than once. In the same respect when calculating reinforcing bars be sure to calculate the exact length of the bars wherever they cross. One of the biggest problems in taking off quantities is that the dimensions on the plans are hard to find and many dimensions are misleading.

EXAMPLE:

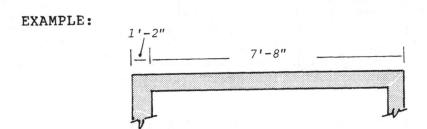

Many candidates will use the 7'-8" dimension as the length and when the calculation is completed the answer will be in the answer band. The 1'-2" dimension could be buried between other dimensions and completely missed.

The best method of becoming familiar with the dimensions and details of the exam plans is to work with old exams with answers that are explained. Much information pertaining to quantity take offs will be found in the notes of the plans. Many times the notes are not on the same page as the details, and/or dimensions.

DESIGN AND CONTROL OF CONCRETE MIXTURES

Concrete as we know it today is composed of portland cement, aggregates and sand, using water as an admixture. Portland cement was produced approximately one hundred fifty years ago. It creates durability and strength when properly mixed with the aggregates and sand. The concrete mixture is produced in a liquid state and hardens readily, usually in a few hours time. Once the concrete mixture begins to harden it will gain strength and moisture. The average hardening time for strength is approximately twenty-eight days.

The book DESIGN AND CONTROL OF CONCRETE MIXTURES describes in detail the actions and reactions of portland cement and how concrete is produced. The test writers lean heavily on this book in the second day of the examination for all General Contractor category's. This book must be understood from cover to cover, as many questions are asked from the graphs and tables and general information regarding the technology of concrete. While studying from this book there is going to be much terminology that will not be understood. It is recommended that the student develop a separate glossary with pages indexed for rapid access. Remember that in the test room, time is of the essence. It should take between 8 to 10 hours of steady reading time to completely read this book. It is recommended that the book be read in one continuous sitting, a few hours at a time, as it is a difficult book to comprehend. Several colored highlighters should be used for mark-up. Follow along with the TABBING AND HIGHLIGHTING MANUAL, although it is recommended that you read the entire book. Study each "graph" and "table" and make sure you understand where and what to select from each "graph" and "table".

If you refer to the index for a particular item in this book and find it not listed, write the word down. When the word is found in the text return to the index and insert it in alphabetical order with a page number on which it appeared in the text.

Be sure to understand the difference between air-entrained and non-air entrained concrete and how air-entrained concrete is produced.

The table on the following page is a guide to the DESIGN AND CONTROL OF CONCRETE MIXTURES book. Should you find additional information to add to this guide while studying, do so.

CONCRETE DESIGN QUIZ

All of the questions pertaining to this quiz are from DESIGN
AND CONTROL OF CONCRETE MIXTURES.

1. Medium sands will have a 22% increase over dry rodded
    sand with an increase of _____ , added by mass.

        A. 15 lb mixture
        B. fine aggregate
        C. 8% specific volume
        D. 10% moisture

2. If the aggregate temp. is 90°F and the concrete temp. is
    90°F the mixing water will be _____°F.

        A. 55
        B. 60
        C. 68
        D. 78
        E. 82

3. Type 1A concrete will have _____ compressive strength
    than Type 1 in 7 days based on relative compressive
    strength requirements.

        A. 62% less
        B. 64% more
        C. 71% less
        D. 25% more
        E. 20% less

4. If the mixing water is 130°F and the weighted average
    temp. of aggregate and cement is 50°F the concrete temp.
    will be _____ .

        A. 72
        B. 67
        C. 60
        D. 58

5. Air entrained concrete will have between _____ and _____ %
    coarse aggregates.

        A.   8 and 10
        B. 15 and 22
        C. 22 and 31
        D. 31 and 51
        E. 51 and 61

6. If the air temp. is 70°F and the RH is 80% and the concrete temp. is 80°F the rate of evaporation will be _____ at a 17 MPH wind velocity.

    A. .65 lbs per sq ft per hr
    B. .42 lbs per sq ft per hr
    C. .33 lbs per sq ft per hr
    D. .20 lbs per sq ft per hr
    E. .15 lbs per sq ft per hr

7. Sea water containing _____ of Magnesium, would be considered suitable for concrete mixing.

    A. between 3000-2200 ppm
    B. between 2210-1800 ppm
    C. between 1810-1400 ppm
    D. between 260-1410 ppm
    E. sea water will not be used for mixing

8. Non-air entrained concrete with a 3-1/2" slump and 3/4" aggregate will contain _____ more lbs per cu yd of water than air entrained concrete.

    A. 340
    B. 260
    C. 110
    D.   35
    E.   12

9. The difference between air entrained and non-air entrained concrete at 28 days with a water cement ratio of .55 based on typical compression strength test of cylinder is _____ .

    A. between 250-500   psi
    B. between 500-1000 psi
    C. between 1000-1500 psi
    D. between 1500-2000 psi
    E. between 2000-2500 psi

10. The entrapped air in air entrained concrete with a fine aggregate of 35% of total aggregate, will be _____ %

    A. 1.4
    B. 2.5
    C. 3.02
    D. 4.5
    E. 5.5

11. If moist cured concrete has a 100% compressive strength in 28 days, it will have a 125% compressive strength in _____ days.

      A. 50 days
      B. 75 days
      C. 100 days
      D. 135 days
      E. 160 days

12. Acid waters with pH values less than _____ may create handling problems and should be avoided if possible.

      A. 25
      B. 16
      C. 7
      D. 3
      E. 1

13. Aggregates generally average about _____ % of concrete volume.

      A. 20
      B. 50
      C. 70
      D. 85

14. Specifications generally require cement to be within _____ % of mixture design.

      A. 8
      B. 6
      C. 3
      D. 2
      E. 1

15. The maximum recommended slump for beams and reinforced walls, using hand methods such as rodding and spading is_____.

      A. 1
      B. 2
      C. 3
      D. 4
      E. 5

ANSWER SHEET

CONCRETE DESIGN QUIZ

1.	D	Chart p 37
2.	C	Chart p 95
3.	E	Table p 23
4.	B	Chart p 104
5.	D	Table p 7
6.	D	Chart p 99
7.	D	Table p 25
8.	D	Table p 63
9.	C	Chart p 11
10.	A	Chart p 46
11.	D	Chart p 87
12.	D	p 27
13.	C	p 29
14.	E	p 66
15.	E	Table p 59

CONCRETE MASONRY

Concrete masonry blocks have been in existance for over one hundred years. Concrete masonry units come in varied shapes and sizes. The most common concrete unit is the 8" x 8" x 16" concrete block, hollow or solid. There are 112.5 block per 100 sq ft, or 1.125 block per one sq ft of surface area. Normal weight concrete block weighs 125 lbs per sq ft. This information can be found on page 7 of the CONCRETE MASONRY HANDBOOK.

Concrete block is defined by nominal size and modular size, as described on page 9 of the CONCRETE MASONRY HANDBOOK. When determining the amount of block required in a concrete block structure, nominal size is used based on 112.5 block per 100 sq ft. It is important to remember that when finding the area of one wall, such as the south wall, that the thickness of the block must be deducted when the adjacent walls (east and west) are calculated, as you cannot overlap or crisscross and take off the proper quantity of block.

The SOUTH FLORIDA BUILDING CODE governing Dade, Broward, and Palm Beach Counties, requires that a tie beam be poured from solid concrete at the top of the block structure to connect the walls to the roof. The STANDARD BUILDING CODE which is used throughout the balance of the State, does not require a tie beam, instead a bond beam block is used, using two No. 5 bars and then poured, constituting the tie rather than the tie beam. Refer to p 15 of the CONCRETE MASONRY HANDBOOK.

In South Florida most block structures are stuccoed for a completed project, where as in North Florida there is much concrete block work that is done with struck block, become familiar with how to apply struck block.

Concrete blocks are assembled with mortar. The CONCRETE MASONRY HANDBOOK, page 31, describes the different types of mortar. Become familiar with these mortars and their reactions. Also know the definition of efflorescence as described on page 34.

Concrete masonry block walls are subject to wind resistance. The tables on pp 47 and 48 discuss wind resistance for non-load bearing walls. Page 139 discusses wind resistance of walls under construction which are still non-reinforced, and ungrouted. Be sure to know what the test writers are asking and which tables to refer to.

The discussion of U and R values on pp 51 through 56 are most important. See p 147 of this manual for further discussion.

Elements of sound, frequency and decibels, must be understood as the test writers have asked questions pertaining to the elements of sound. The decibel is the most widely means of measurement used by building codes to regulate the amount of noise to be tolerated. The decibel meter measures sound pressure level in microbars. The higher the decibel reading the louder the noise. Average conversation is in the 50 decibel range and most building codes consider a range of from 40 to 55 db as a comfortably unannoying range. See CONCRETE MASONRY HANDBOOK, page 57 and OSHA, Section 1926.52.

Acoustic ratings are identified by frequency or by class. Acoustic ratings are measured in sabins, which are units of energy absorbtion. The higher the rating the better the sound absorbtion (known as diaper effect). This is shown in the CONCRETE MASONRY HANDBOOK, page 60, Table 3-13. On page 59 a noise reduction table, Table 3-12, is shown. Understand both of these tables as questions have been asked regarding them.

Design and Layout of Concrete Masonry Walls:

Concrete masonry walls require reinforcing. The cavity wall (See page 73 CONCRETE MASONRY HANDBOOK), requires wall ties which are 4" x 8" ties. Composite walls use bonding courses every 7th course. Reinforced walls use steel reinforcement. Joint reinforcement is used as horizontal reinforcement to control cracking and to bond the wythes without using metal ties. The test writers refer to joint reinforcement as H.J.R'S. See page 79, CONCRETE MASONRY HANDBOOK. Also understand control joints as described on page 89, CONCRETE MASONRY HANDBOOK.

Much concrete block work is done as struck block, where the exterior of the block is not stuccoed. To do this type of masonry the block must be laid in stretchers. (See pages 95-97, CONCRETE MASONRY HANDBOOK.) This is also known as modular planning, where all horizontal dimensions are in full block (16") or half block (8"). Vertical dimensions are in full height (8") or in half block (4"). Understand door and window opening applications and how to apply them properly.

Test questions have pertained to volume of mortar required to stucco an area of masonry structure. Normally, approximately 1/2" of thickness is needed. Remember 1/2" is equal to 0.042 ft and not .5 to determine volume.

Read and study the CONCRETE MASONRY BOOK well; many questions have been asked from it.

PLACING REINFORCING BARS

Concrete is strong in compression but weak in tension and shear; therefore, reinforcement must be incorporated in the concrete in order to attain the strength for tension and shear. Reinforcement is usually deformed bars varying from 3/8" to 2-1/2" in diameter. Welded wire fabric is also used either in place of or in combination with deformed bars, see WALKER'S ESTIMATOR'S REFERENCE BOOK, page 8.89, for characteristics of wire fabric.

Concrete is the strength of the compression factor and steel is the function of tension or pull resistance. The deformation or lugs on the steel are the bonding agents for the concrete. See Chapter 3, of PLACING REINFORCING BARS, for various types of concrete placement.

Page 5-1 describes the compressive strength of concrete and how compression tests are conducted. Questions have been asked off this page. Page 5-2 describes vertical and horizontal shear. Pages 5-3 through 5-8 describe the most efficient placement of steel in concrete structures.

Page 6-8 defines bar size and bar weight of standard reinforcing bars. Know this table thoroughly. Pages 6-5 through 6-16 describe bar bends. Pages 7-5 through 7-9 discuss hoisting of reinforcing bars. Many questions have been asked of the diagrams, tables, and formulas. Know how to apply the table for Dropped Forged Hoist Hooks on page 7-7 and the formula for lifting bars on page 7-9.

Page 9-2 describes wire bar supports, become familiar with each detail on this page, and the entire chapter.

Be aware that reinforcing bar may be welded as shown in Chapter 10. The reinforcing bars must be overlapped. The acceptable practice is to overlap 16" to 18". The test writers refer to overlap in terms of bar diameters. For example; using a 5/8" bar, the overlap is to be 24 bar diameters. Multiplying .625 (the diameter of one bar) by 24, (.625 x 24) = 15 inches.

Pages 11-19 and 11-21 show column ties. The ties shown on page 11-19 are "UNIVERSAL STANDARD COLUMN TIES" and the ones shown on page 11-21 are "ACI STANDARD CLOSED COLUMN TIES". Know these two pages and the difference between them.

Study Chapter 12 in its entirety; joist construction is very important.

STEEL CONSTRUCTION

Referenced page numbers in this chapter are to the STEEL
CONSTRUCTION MANUAL, 8th Edition, published by the American
Steel Institute.

Structural steel for building construction is available in
various shapes and sizes. It is important to understand the
symbols and nomenclature for each of the various types.

STRUCTURAL SHAPES

        Structural shapes are primarily:

                        W Shapes
                        M Shapes
                        S Shapes
                        HP Shapes
                American Standard Channels (C)
                Miscellaneous channels (MC)
                        Angles (L)

W, M, S, HP, C, MC and L Shapes are designated by their
letter which is followed by two numbers. The first number is
the depth (d) dimension and the second number is the weight
per running foot. For example a W18 x 71 is 18 inches deep
and weighs 71 pounds per running foot. Refer to pages 1-14
through 1-35 for dimensions and properties.

AMERICAN STANDARD CHANNELS (C) are designated by a letter
followed by two numbers, the same as HP shapes, with the
numbers referring to depth and weight. A reference given to
channels is grip (see definition, page 5-221, 2nd paragraph).
Refer to pages 1-36 and 1-37 for dimensions and properties.

MISCELLANEOUS CHANNELS (MC) are designated the same as (C)
channels. Refer to pages 1-38 through 1-41 for dimensions
and properties.

ANGLES (L) are designated by letter followed by three
numbers. The first two numbers are the length of both legs
known as the X and Y axis. The third number is the thickness
of the steel. The weight will be found in column 3. For
properties see pages 1-42 through 1-46.

STRUCTURAL TIES which are made by cutting W, M, and S shapes
in half, are designated by the letter followed by two numbers
the same as W, M and S shapes. Refer to pages 1-50 through
1-69.

DOUBLE ANGLES are two angles joined together and are
designated the same as angles. Refer to pages 1-72 through
1-77 for dimensions and properties.

COMBINATION SECTIONS are the joining of shapes and angles for special applications. Refer to pages 1-80 through 1-87 for dimensions and properties.

STEEL PIPE is listed as Standard Weight (schedule 40) Extra Strong (schedule 80) and Double Extra Strong (schedule 120) on page 1-89. Steel pipe is designated by nominal inside dimension. As the pipe gets heavier the inside dimension gets smaller and the weight per foot is heavier. The reason for working with inside dimensions is that standard fittings are used for all three thicknesses.

STRUCTURAL TUBING (square and rectangular) are designated by an X and Y axis which refers to the outside dimensions. Each size is also designated by wall thickness. Refer to pages 1-90 through 1-96 for dimensions and properties. Both round pipe and structural tubing are used in construction as column supports, referred to as pipe columns and lolly columns. The difference being that lolly columns are pipe columns filled with concrete. Lolly columns are used in structures at critical locations where fire could weaken, bend or melt the column thus the concrete in the column prolongs the durability of the column. The Florida Contractors Exam asks questions regarding the volume of concrete required in a column or a number of columns. To find the area of round columns use the inside dimension from page 1-89. For structural tubing the outside dimension is given; therefore, take the area of the outside dimension in square inches then subtract the area of the thickness of the tubing to determine inside area.

FOR EXAMPLE:

From page 1-92, a 14 x 10 x 3/8 structural tube has an area of 14 x 10 = 140 sq inches less (from column 5) area of tube thickness of 17.1 sq in = 122.9 sq inches divided by 144 sq inches per sq ft = 122.9/144 = .85 sq ft of area.

BARS AND PLATES

Steel plates are designated as flat plate (or base plate) and floor plate. Floor plate is heavier than flat plate due to raised patterns designed for traction. One square foot of floor plate one inch thick weighs 41.89 lbs; whereas, the same amount of flat plate (base plate) weighs 40.8 lbs per sq ft, (See WALKER'S ESTIMATORS page 10.21). Flat plate (base plate) is used as the base of steel columns. Refer to page 1-102 and 1-103 for weights and thicknesses.

CRANE RAIL

It is important to note that the weight of crane rail is specified in yards and not feet. Standard crane rail weighs 85 lbs per linear yard. Refer to page 1-105 for weights and

specifications. Also refer to page 1-106 for welded and
bolted splices. Refer to the last paragraph on page 1-106
for maintenance.

Test questions are asked regarding the painting, sealing, or
covering of the surfaces of W shape steel. Refer to page
1-117 through 1-119 for tables of surface areas per linear ft
of W shape. Study and understand this page then highlight
the surfaces referenced in cases A, B, C, and D. Questions
are usually asked about quantities of paint needed for a
particular amount of W Shape.

Pages 1-122 through 1-124 describe camber and sweep of
various shapes. Camber is the bending of the member along
the base, as in beams. It can be camber up, to anticipate
load, or camber down due to overload. Sweep is the bending
in the vertical plane which would indicate an out of plumb
condition. Also refer to page 6-6 for use of heat in
straightening camber.

UNIFORM LOAD CALCULATIONS

For W shapes based on Fy - 36 KSI, and Fy - 50 KSI, Fy is the
specified minimum yield stress in KSI (Kips per square inch).
One Kip = 1000 lbs. Refer to page 2-24 and 2-26 for uniform
load capacities.

Examples 1 and 2: Example 1 is for the Lv (see nomenclature
page 2-24 for definitions of Lv) when the Lv is less than the
length of the beam selected. Example 2 is for beams where
the beam selected is less than the Lv.

Example 3 is for three concentrated loads. When using these
formulas make sure information is selected from the proper
tables (pages 2-27 through 2-45 for Fy = 36 KSI or Fy = 50
KSI. Questions have been asked about Examples 1 and 2.

BOLTS AND NUTS

Many second day questions have been referenced to bolts and
nuts. Refer to pages 4-135 through 4-137 for symbols and
definitions. Page 4-138 and 4-140 should be studied and
Understood for weight of bolts, nuts, and washers. High
strength bolts ASTM A325 and A490 are most important to
understand for test purposes. Study pages 5-209 through
5-220 for a basic coverage of ASTM 325 and A490 bolts.
Understand Table G - page 5-221 for grip.

WELDING

Welding is covered in three different books, STEEL MANUAL
pp 4-146 thru 4-165, WALKER'S ESTIMATORS pp 10.9 thru 10.14
and the BUILDING TRADES DICTIONARY under W. It is
recommended that the welding sections of all three books be
studied.

BUSINESS AND FINANCIAL MANAGEMENT

Examination questions on business and management may vary greatly. They may be limited to excerpts from local statutes and codes dealing with legal financial requirements, insurance and licensing, or they may include general, management, reading an accountant's statement and/or cost estimating and job pricing. Whatever these questions deal with, we may expect more questions on business and management in forthcoming exams than ever before.

THE LEGALITY OF BUSINESS

The principal forms of business organization that may be used for a mechanical contracting business are (1) single proprietorship, (2) part partnership, and (3) corporation. Each has advantages and disadvantages. Because of limited financial liability for stockholders, the corporation is the most common form of organization among contractors. Freedom of operation and tax angles are other things to be considered. Your decision about which to use should be made only after careful study and consultant and attorney.

MANAGEMENT FUNCTIONS

Each year, several thousand businesses fail. According to the Small Business Administration, nine out of every ten failures are caused by poor management. Running a business requires more than knowing how to sweat a joint. Analysts regard the following as being the important areas in which poor management finally results in business failure:

Unbalanced experience and inability to;
1. Control operating expenses
2. Grant credit intelligently
3. Management finances
4. Price jobs correctly

It is not uncommon to hear an ex-contractor blame his failure on "the competition", "a crooked bookkeeper", "lousy help", "cheap customers", but in the final analysis it is he himself who failed at his management function.

The management functions are:

1. General administration
2. Legal matters
3. Personnel
4. Design and engineering.
5. Construction
6. Purchasing and subcontracting
7. Financial control
8. Planning
9. Sales
10. Service

WHAT FORM OF BUSINESS ORGANIZATION?

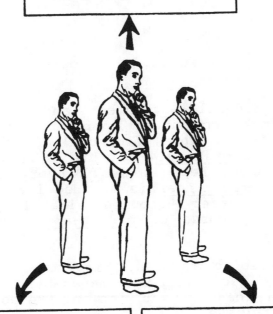

Individual Proprietorship

One person as sole owner.
No legal organization requirements.
Unlimited personal liability for business debts.
Termination upon death of owner.
Relative freedom from Government control.
No income tax levy on business (but on owner only).

Partnership

Two or more persons as co-owners.
Definition of partner's rights and duties by partnership agreement and State partnership law.
Possible requirement for filing certificate under State law.
Unlimited liability of each partner for all debts of firm.
Capacity of one partner to bind others when acting within scope of business.
Termination upon death or withdrawal of a partner.
Relative freedom from government control.
No income tax levy on partnership itself.

Corporation

Creation as a legal entity pursuant to State law.
Stockholders as owners who are separate and distinct from corporation.
Requirement for obtaining charter from State.
Required payment of filing fees and capital stock taxes.
Continuity unaffected by death or transfer of stock shares by any or all owners.
Limited financial liability for stockholders.
Subjection to more Government control than a proprietorship or partnership.
Income tax levy upon corporate profits and, in addition, upon dividends after they are paid to stockholders.

FIGURE 1

Scheduling and coordinating the construction department is among the most important overall responsibilities. Planning means keeping the company's resources occupied constantly without ever overexpanding beyond those resources. Those resources are obviously, manpower, equipment and finances.

When a company's sales order backlog is below what is required to fully utilize its manpower, equipment and finances, soft pricing is needed. Conversely, the higher the sales order backlog, the firmer the policy. Soft pricing means: price concessions, liberalized credit policy and terms, special discounts, dynamic sales promotions, and to other inducements such as early delivery, extended warranties and attractive service programs.

Record Keeping

No contracting business can operate profitably without accurate record keeping.

First, the Federal Government and most States require that businesses make up annual operating and financial statements for tax purposes. Then, financial control of a contracting business requires a cash flow picture. An appropriate journal or cash book must be kept in which every in-and-out item is recorded. The general ledger is divided into different accounts that may include: asset accounts, liability accounts. This information is the source of profit-and-loss statements; (P&L and balance sheets.

More than keeping records, they must be used. Constant study of company records gives management a "feel" for the business , which leads to an instant response to variations that is almost intuitive.

Job cost analysis--with accurate feedback--labor time reports, weekly payroll records, wage report and employees, are all part of the record keeping routine and the responsibility of management.

HOW TO READ A FINANCIAL STATEMENT

Profit-and-loss statements prepared by accounts show not only dollar totals, but usually also the percentage of sales represented by each item. Percentages, of course, are expressions of arithmetical proportions. Proportions are ratios.

There are three kinds of ratios. The first are balance sheet ratios that refer to relationships between various balance sheet items. The second are the operating ratios that show the relationship of expense accounts to income. The third group is made up of ratios that show the relationship between an item in the P&L statement and one on the balance sheet.

A balance sheet tells how a business stands at one given moment in the business year. A profit-and-loss statement sums up the results of operations over a period of time.

Of themselves, these two types of of financial documents are a collection of inanimate figures. But when the assorted financial symbols are interpreted and evaluated, they begin to talk.

A single balance sheet is like the opening chapters of a book--it gives the initial setting. Thus, one balance sheet will show how the capital is distributed, how much is in the various accounts, and how much surplus of assets over liabilities exists. A lone profit-and-loss statement indicates the sales volume for a given period, the amount of costs incurred, and the amount earned after allowing for all costs.

When a series of balance sheets for regularly related intervals, such as fiscal or calendar yearends, is arranged in vertical columns so that related items may be compared, the changes in these items begin to disclose trends. The comparative balance sheets then no longer remain snapshots, but are converted into X-rays, penetrating outward tissue and outlining skeletal structure of all basic management actions and decisions.

Thus, decisions to increase basic inventories because of upward price changes may be revealed in larger quantities of merchandise on hand from one period to the next. If credits are relaxed and collections slow up when sales remain constant, there may be a successive increase in receivables. If expansion is undertaken, debts may run higher; and if losses are sustained, net worth declines.

Similarly, comparative profit-and-loss statements reveal significant changes in what took place. Were prices cut to meet competition? Then look for a lower gross profit--unless purchasing costs were reduced proportionately. Did sales go up? If so, what about expenses? Did they remain proportionate? Was more money spent on office help? Where did the money come from? How about fixed overhead? Was it controlled? It is only by comparing operating income and cost account items from one period to another that revealing answers are found.

Ratios - A Physical Exam to Determine a Company's Health

The following 10 ratios are suggested as key ones for small business purposes:

1. Current assets to current liabilities.
2. Current liabilities to tangible net worth.
3. Turnover of tangible net worth.
4. Turnover of working capital.

5. Net profits to tangible net worth.
6. Average collection period of receivables.
7. Net sales to inventory.
8. Net fixed assets to tangible net worth.
9. Total debt to tangible net worth.
10. Net profit on net sales.

Brief definitions of these ratios appear below, followed by specific examples using data taken from the balance sheet and profit-and-loss statement seen on Figures 2 and 3. Explanation of the terms of the financial statements, used in calculating the ratio, is also included in the discussion of each ratio.

1. <u>Current assets to current liabilities.</u> Widely known as the "current ratio," this is one test of solvency, measuring the liquid assets available to meet all debts falling due within a year's time.

$$\text{Example:}\quad \frac{\text{Current assets}}{\text{Current liabilities}} = \frac{\$37,867}{\$19,242} = 1.97 \text{ times.}$$

Current assets are those normally expected to flow into cash in the course of a merchandising cycle. Ordinarily these include cash, notes and accounts receivable, and inventory, and at times, in addition, short term and marketable securities listed on leading exchanges at current realizable values. While some concerns may consider current items such as cash-surrender value of life insurance as current, the tendency is to post the latter as noncurrent.

Current liabilities are short term obligations for the payment of cash due on demand or within a year. Such liabilities ordinarily include notes and accounts payable for merchandise, open loans payable, short term bank loans, taxes, and accruals. Other sundry short term obligations, such as maturing equipment obligations and the like, also fall within the category of current liabilities.

2. <u>Current liabilities to tangible net worth.</u> Like the "current ratio," this is another means of evaluating financial condition by comparing what is owed to what is owned. If this ratio exceeds 80 percent, this is considered a danger sign.

$$\text{Example:}\quad \frac{\text{Current liabilities}}{\text{Tangible net worth}} = \frac{\$19,242}{\$33,970} = 56.6 \text{ percent}$$

Tangible net worth is the worth of a business, minus any tangible items in the assets such as goodwill, trademarks, patents, copyrights, leaseholds, treasury stock, organization expenses, or underwriting discounts and expenses. In a corporporation , the tangible net worth would consist of the sum of all outstanding capital stock--preferred and common--and surplus, minus intangibles. In a partnership or proprietorship, it could comprise the capital account, or accounts, less the intangibles.

A word about "intangibles." In a going business, these items frequently have a great but undeterminable realizable value. Until these intangibles are actually liquidated by sale, it is difficult for an analyst to evaluate what they might bring. In some cases, they have no commercial value except to those who hold them: for instance, an item of goodwill. To a profitable business up for sale, the goodwill conceivably could represent the potential earning power over a period of years, and actually bring more than the assets themselves. On the other hand, another business might find itself unable to realize anything at all on goodwill.

3. <u>Turnover of tangible net worth.</u> Sometimes called "net sales to tangible net worth," this ratio shows how actively invested capital is being put to work by indicating its turnover during a period. It helps measure the profitability of the investment. Both overwork and underwork of tangible net worth are considered unhealthy.

$$\text{Example: } \frac{\text{Net sale}}{\text{tangible net worth}} = \frac{\$189,754}{\$33,970} = 5.6 \text{ times}$$

Turnover of tangible net worth is determined by dividing the average tangible net worth into net sales for the same periods. The ratio is expressed as the number of times the turnover is obtained within the given period.

4. <u>Turnover of working capital.</u> Known, as well, as the ratio of "net sales to net working capital," this ratio also put to work in terms of sales. Working capital or cash is assets that can readily be converted into operating funds within a year. It does not include invested capital. A low ratio shows unprofitable use of working capital; a high one, vulnerability to creditors.

$$\text{Example: } \frac{\text{Net sales}}{\text{working capital}} = \frac{\text{net sales}}{\text{current assets-current liabilities}}$$

$$= \frac{\$189,754}{\$37,867 - \$19,242} = 10.2 \text{ times.}$$

Deduct the sum of the current liabilities from the total current assets to get working capital, the business assets which can readily be converted into operating funds. A business with $100,000 in cash, receivables and inventories and no unpaid obligations would have $100,000 in working capital. A business with $200,000 in current assets and $100,000 in current liabilities also would have $100,000 working capital. Obviously, however, items like receivables and inventories cannot usually be liquidated overnight. Hence, most businesses require a margin of current assets over and above current liabilities to provide for stock and work-in-process inventory, and also to carry ensuing receivables after the goods are sold until the receivables are collected.

5. <u>Net profits to tangible net worth.</u> As the measure of return on investment, this is increasingly considered one of the best criteria of profitability, often the key measure of management efficiency. Profits "after taxes" are widely looked upon as the final source of payment on investment plus a source of funds available for future growth. If this "return on capital" is too low, the capital involved could be better used elsewhere.

$$\text{Example:} \quad \frac{\text{Net profits (after taxes)}}{\text{Tangible net worth}} = \frac{\$5,942}{\$33,970} = 17.5 \text{ percent}$$

This ratio relates profits actually earned in a given length of time to the average net worth during that time. Profit here means the revenue left over from sales income and allowing for payment of all costs. These include costs of goods sold, writedowns and chargeoffs, Federal and other taxes accruing over the period covered, and whatever miscellaneous adjustments may be necessary to reduce assets to current, going values. The ratio is determine by dividing tangible net worth at a given period into net profits for a given period. The ratio is expressed as a percentage.

6. <u>Average collection period of receivable.</u> This ratio, known also as the "collection period" ratio, shows how long the money in a business is tied up in credit sales. In comparing this figure with net maturity in selling terms, many consider a collection period excessive if it is more than 10 to 15 days longer than those stated in selling terms. To get the collection period figure, get average daily credit sales, then divide into the sum of notes and accounts receivable.

$$\text{Example:} \quad \frac{\text{Net (credit) sales for year}}{365 \text{ days a year}} = \text{daily (credit) sales (\$519)}$$

$$\text{Average collection period} = \frac{\text{notes and accounts receivable}}{\text{daily (credit) sales}}$$

$$= \frac{\$26,765}{\$519} = 51.6$$

This figure represents the number of days' sales tied up in trade accounts and notes receivable or the average collection received. The receivables discounted or assigned with recourse are included because they must be collected either directly by borrower, or by lender; if uncollected, they must be replaced by cash or substitute collateral. A pledge with recourse makes the borrower just as responsible for collection as though the receivables had not been assigned or discounted. Aside from this, the likely collectibility of all receivables must be analyzed, regardless of whether or not they are discounted. Hence all receivables are included in determining the average collection period.

7. <u>Net sales to inventory</u>. Known also as a "stock-to-sales" ratio, the hypothetical "average" inventory turnover figure is valued for purposes of comparing one company's performance with another, or with the industry's.

$$\text{Example:} \quad \frac{\text{Net sales}}{\text{Inventory}} = \frac{\$189.754}{\$10,385} = 18.3 \text{ times.}$$

A manufacturer's inventory is the sum of finished merchandise on hand, raw material, and material in process. It does not include supplies unless they are for sale. For retailers and wholesalers, it is simply the stock of salable goods on hand. It is expected that inventory will be valued conservatively on the basis of standard accounting methods of valuation, such as its cost or its market value, whichever is the lower.

Divide the average inventory into the net sales over a given period. This shows the number of times the inventory turned over in the period selected. It is compiled purely from one period to another, or for other comparative purposes. This ratio is not an indicator of physical turnover. The only accurate way to obtain a physical turnover figure is to count each type of item in stock and compare it with the actual physical sales of that particular item.

Some people compute turnover by dividing the average inventory value at cost into the cost of goods sold for a particular period. However, this method still gives only an average. A hardware store stocking some 10,000 items might divide its dollar inventory total into cost of goods would hardly define the actual turnover of each item from paints to electrical supplies.

8. <u>Fixed assets to tangible net worth</u>. This ratio, which shows the relationship between investment in plant and equipment and the owner's capital, indicates how liquid is net worth. The higher this ratio, the less the owner's capital is available for use as working capital, or to meet debts.

$$\text{Example:} \quad \frac{\text{Fixed assets}}{\text{Tangible net worth}} = \frac{\$15.345}{\$33,970} = 45.2 \text{ percent}$$

```
                    STATEWIDE CONSTRUCTION CO.

                          Balance Sheet

                        December 31, 19__


                             Assets

Current assets:
  Cash on hand and in banks. . . . . . . . . . . .     $ 4,320
  Notes receivable . . . . . . . . . . . . $ 4,820
    Less notes discounted. . . . . . . . .   3,000      1,820

  Accounts receivable. . . . . . . . . .    21,945
    Less reserve for bad debts . . . . . .   1,875     20,070

  Inventories. . . . . . . . . . . . . . . . . . .     10,385
  Prepayment of expenses . . . . . . . . . . . . .      1,272

        Total current assets. . . . . . . . . . .     $37,867

Plant and equipment:
  Land and building. . . . . . . . . . . . $14,495
  Equipment, fixtures, and furniture . . .   4,800
    Less allowances for depreciation . . .   3,950     15,345

Intangibles:
  Goodwill . . . . . . . . . . . . . . . .     500
  Patent, franchises, etc. . . . . . . . .     500      1,000

        Total assets. . . . . . . . . . . .           $54,212


                           Liabilities

Current liabilities:
  Notes Payable (to banks) . . . . . . . . . . . .     $ 4,000
  Accounts Payable (trade) . . . . . . . . . . . .      10,322
  Taxes payable. . . . . . . . . . . . . . . . . .       3,600
  Other payables (including accruals). . . . . . .       1,320

        Total current liabilities . . . . . . . .     $19,242
Fixed liabilities. . . . . . . . . . . . . . . . .           0

        Total liabilities . . . . . . . . . . . .     $19,242


                            Capital

Capital stock (preferred). . . . . . . . . $ 5,000
Capital stock (common) . . . . . . . . . .  20,000
Surplus. . . . . . . . . . . . . . . . . .   9,970

        Total . . . . . . . . . . . . . . . . . .     $34,970

        Total liabilities and capital . . . . . .     $54,212
```

FIGURE   2

STATEWIDE CONSTRUCTION CO.

Profit and Loss Statement

For year ending December 31, 19___

Item	Amount	Percent
Gross sales.	.$193,472	
Less returns and allowances. . .	3,718	
Net Sales.	.$189,754	100.00
Cost of goods sold	. 147,348	77.65
Gross profit on sales.	.$ 42,406	22.35
Selling expenses . . . .$ 10,479		5.52
General and administrative exp . 19,510		10.28
Financial expenses 1,312		.69
Total expenses	. 31,301	16.49
Operating profit	.$ 11,105	5.86
Extraordinary expenses	. 300	.16
Net profit before taxes.	.$ 10,805	5.70
Federal, State, and local taxes.	. 4,863	2.56
Net profit after taxes	.$ 5,942	3.14

FIGURE 3

Financial expenses. This item would include interest, doubtful accounts, and discounts granted if not already deducted from sales.

Other operating expenses and income. Here might be included various unusual expense items not elsewhere classified, such as moving expenses, against which might be credited income from investments and miscellaneous credits and debits.

Extraordinary charges (if any). Such expenses do not occur very often, but occasionally unusual costs such as losses on sale of unused fixtures and equipment do arise.

Net profit before taxes. This figure is the profit after deducted the regular and extraordinary business charges mentioned above.

Taxes. This item includes the Federal, State and local taxes paid by the company out of its earnings.

of each item to a common base of net sales. These percentages may be compared with those of previous periods to measure a firm's performance. They also may be compared to the typical percentages of business in similar trades or industries when they are available. Such comparisons will indicate the competitive strengths and weaknesses of a business.

The items included in profit and loss statements vary from business to business. For example, some business break down their sales expense to show the cost of salesmen's salaries and commissions, advertising, delivery costs, supplies,and so forth; some do not. In the following explanation of the P & L items, only major items are included.

The following explanations briefly discuss each term in the accompanying condensed profit-and-loss statement.

Net sales. This figure represents gross dollar sales minus merchandise returns and allowances. Some accountants also deduct cash discounts granted to customers on the theory that these are actually a reduction of the net selling price; "quantity" discounts are, of course, concessions off price, and should be deducted from the gross sales. In setting up the profit-and-loss statement in percentages, the net sales are shown as 100 percent.

Cost of goods sold. For retailers and wholesalers, this figure is the inventory at the beginning, plush purchases, plus "Freight in," and less inventory at the end of the period."Freight out" is generally shown as delivery expense, either under separate or other sections of the statement.

For manufacturers, there are various additional items to be considered. They include supervision, power, supplies, the direct costs of manufacturing labor (including social security and unemployment taxes on factory employees), that portion of depreciation which enters into cost of production and many others.

Gross profit on sales. This figure is obtained by deducting cost-of-goods sold from net sales.

Selling expenses. These expenses include such items as salaries of salesmen and sales executives, wages of other salesemployees, commissions, travel expense, entertainment expense, and advertising.

Operating profit. This is the difference between the gross profit on sales and the sum of the selling expenses.

General and administrative expenses. These expenses include officers' salaries, office overhead, light, heat,communication, salaries of general office and clerical help, cost of legal and accounting services, "fringe" taxes payable on

Fixed assets means the sum of assets such as land, buildings, leasehold improvements, fixtures, furniture, machinery, tools, and equipment, less depreciation. The ratio is obtained by dividing the depreciated fixed assets by the tangible net worth.

9. <u>Total debt to tangible net worth.</u> This ratio also measures "what is owed to what is owned." As this figure approaches 100, the creditors' interest in the business assets approaches the owner's.

$$\text{Example:} \quad \frac{\text{Total debt}}{\text{Tangible net worth}} = \frac{\text{current debt+ fixed debt}}{\text{tangible net worth}}$$

$$\frac{\$19,242}{\$33,970} = 56.6$$

Total debt is the sum of all obligations owed by the company such as accounts and notes payable, bonds outstanding, and mortgage payable. The ratio is obtained by dividing the total of these debts by tangible net worth.

10. <u>Net profit on net sales.</u> This ratio measures the rate of return on net sales. The resultant percentage indicates the number of cents of each sales dollar remaining, after considering all income statement atement items and excluding income taxes.

A slight variation of the above occurs when net operating profit is divided by net sales. This ratio reveals the profitableness of sales--i.e., the profitableness of the regular buying, manufacturing, and selling operations of a business.

To many, a high rate of return on net sales is necessary for successful operation. This view is not always sound. To evaluate properly the significance of the ratio, consideration should be given to such factors as (1) the value of sales (2) the total capital employed and (3) the turnover of inventories and receivables. A low rate of return compared with rapid turnover and large sales volume, for example, may result in satisfactory earnings.

$$\text{Example:} \quad \frac{\text{Net profits}}{\text{net sales}} = \frac{\$5,942}{\$189,754} = 3.1 \text{ percent}$$

Analyzing the Profit-and-Loss (Income) Statement

Based solely on data taken from the profit-and-loss (P&L) statement, operating ratios show the percentage relationships administrative personnel, sundry types of franchise and similar taxes, and other expenses.

Financial expenses. This item would include interest, doubtful accounts, and discounts granted if not already deducted from sales.

Other operating expenses and income. Here might be included various unusual expense unusual expense items not elsewhere classified, such as moving expenses, against which might be credited income from investments and miscellaneous credits and debits.

Extraordinary charges (if any). Such expenses do not occur very often, but occasionally unusual costs such as losses on sale of unused fixtures and equipment do arise.

Net profit before taxes. This figure is the profit after deducted from the regular and extraordinary business charges mentioned above.

Taxes. This item includes the Federal, State and local taxes paid by the company out of its earnings.

Net profit after taxes. This figure is the final figure showing earnings available for distribution or retention.

Figure 3 illustrates how a condensed profit-and-loss statement would be expressed, first in terms of dollars, then in terms of percentages of net sales.

Here are some established average ratios for air conditioning, plumbing and mechanical contractors. Companies within this range may be considered healthy operations.

1. Current assets to current liabilities   2 times
2. Net profit on net sales   1.5 percent
3. Net profit on tangible net worth   10 percent
4. Net sales to tangible net worth   7.5 percent
5. Current liabilities to tangible net worth   100 percent
6. Total disabilities   150 percent
7. Net profits on net working capital   15 percent

MONEY AND INTEREST

For the past several years, construction contractors have accounted for a disproportionate number of business failures in this country. For example, during a recent year in which construction accounted for 8 percent of the gross national product, contractors accounted for approximately 20 percent of all financial failures and 23 percent of the resulting liabilities. Studies by Dun & Bradstreet, Inc. disclose that the underlying management weaknesses in descending order of importance are as follows:

1. Incompetence
2. Lack of company expertise in sales, financing and purchasing
3. Lack of managerial experience
4. Lack of experience in firm's line of work
5. Fraud, neglect, disaster, or unknown

Consideration of these factors makes it clear that the financial success of a construction enterprise depends almost entirely on the quality of its management. Many outward manifestations of these root causes can be given for business failure: inadequate sales, competition weaknessess, heavy operating expenses, low profit margins, cash flow difficulties resulting from slow; payments, and over extension. It is obvious, however, that these business inadequacies are simply indicative of poor business management.

Because of this situation, the Florida Department of Professional Regulation through its regulatory body, The Construction Industry Licensing Board decided to address the issue of the lack of management expertise by requiring all candidates for license to have a basic understanding of management; principles and practices as exemplified by passing on examination on these issues.

As you will notice from the Candidate Information Booklet beginning on page 9 , Part I of the examination covers Business and Financial Management. It gives a detailed description of each of the items to be emphasized on the exam and their relation percentage of the exam content.

It has been the practice of the Construction Industry Licensing Board to require a detailed understanding of the following topical areas in order to obtain a passing grade on the Business and Financial Management section of the state exam for license:

    Chart of Accounts
    Net Worth
    Reconciling Bank Statements
    Petty Cash Funds

Aging of Accounts Receivable
Cost at Completion
Completed Contract Method of Accounting
Breakeven Analyses
Forecasting of Sales & Profits
Efficiency of Work Crews
Allocation of Overhead by Job
Types of Depreciation
Interest of Loans
Financial Ratios
Types of Insurance
Calculation of Insurance Premium
Cost Estimating
Overtime Compensation
Payroll Taxes
Unemployment Compensation
Workers Compensation
Contractual Obligations
Applications for Partial Payment

The reference book, BUILDER'S GUIDE TO ACCOUNTING by Michael C. Thomsett, 2nd edition, 1987, Craftsman Book Company, Carlsbad, CA, is a well written, comprehensive book on the subject of accounting. It is used as a major instruction tool for the Contractor's exam. For example, when addressing the topic of the Chart of Accounts one should refer to pages 257 and 262, and all of Chapter 1 (pages 5-16) in the aforementioned publication. Chapters 1 and 21 in BUILDER'S GUIDE TO ACCOUNTING will assist the reader in understanding the concept of Net Worth and those items that affect it. Reconciling Bank Statements can be found in Chapters 9 and 18.

Petty Cash funds are described in Chapter 17 on pages 174 with a sample format shown on page 288. Chapter 2 covers the Methods of Accounting, while Aging of Receivables is capabaly explained in Chapter 4. Each of the remaining chapters of the BUILDERS GUIDE TO ACCOUNTING covers the material necessary to prepare the reader to become more proficient in the areas of business and financial management.

The following **Description of Compound Interest Factors** and quizzes will help the student become familiar with the business requirements of the use of money and how the value of money is related to time and the rate of interest. The page references in the Description of Compound Interest Factors are made to Chapter 10 of WALKERS PRACTICAL ACCOUNTING. The compound interest tables at the back of WALKERS PRACTICAL ACCOUNTING range from 5% to 25% in increments of one-quarter of one percent and show a single payment schedule and an equal annual payment schedule. How to use these tabes is explained in the Description of Compound Interest Factors.

DESCRIPTION OF COMPOUND INTEREST FACTORS

Compound Interest Tables of various percentages begin on page 152. They start at 5% and go up 25%. The factors in each table are the same except for the interest rate.

The first column, N is the number of years.

SINGLE PAYMENT

The first column, **P to find F**, the amount of money you would have at the end of N years with one initial investment. $100.00 today at 10% for 5 years is equal to F ($161.05).

The third column, **F to find P**, is what money in the future would be worth today. If you were to receive a $1000 payment two years from now, what would it be worth today? (p 172) $1000 at 10% received two years from now would be worth $1000 x .82645 or $826.45 today.

UNIFORM ANNUAL SERIES

The fourth column, **F to find A**. In order to receive $10,000 at the end of 5 years how much must you deposit at the end of each year at 10% (p 172) F = $10,000 at the end of 5 years. A = How much each year? $10,000 x .163797 = $1637.97.

The fifth column, **P to find A**. What annual payments would you make per year for N years when you know the amount you want to borrow? If you want to borrow $10,000 payable over 5 years at 10% interest what would your annual payment be? (page 172) $10,000 x .263797 = $2637.97.

The sixth column, **A to find F**. If you invested an equal amount each year at a given rate of interest what would you have at the end of N years, if you deposited $1000 per year for 5 years at 10%? (page 172) $1000 x 6.1051 = $6,105.10

The seventh column, **A to find P**. If you know how much you can pay per year for N years at the rate of interest how much can you borrow or receive today? (page 173) $1000 x 2.7908 = $3790.80.

BALLOON MORTGAGE

A balloon mortgage is based on selecting a rate of interest for a given period of time but completing the transaction in a lesser time with a lump sum payoff at the end. You may, for example, borrow $10,000 at a rate of 18% interest for a period N = 15 years, but to balloon at the end of five years. As shown in the example pp 148-149, the annual payments for the first four years would each be $1964.03 and the final 5th year payment would be $10,790.51, for a total payoff of $18,646.63.

TABLE 1: 6% COMPOUND INTEREST FACTORS

1	2	3	4	5	6	7	8
	Single payment		Uniform series				
	Compound amount factor	Present worth factor	Sinking fund factor	Capital recovery factor	Compound amount factor	Present worth factor	
N	Given P to find F $P \xrightarrow{i} F, n=$	Given F to find P $F \xrightarrow{i} P, n=$	Given F to find A $F \xrightarrow{i} A, n=$	Given P to find A $P \xrightarrow{i} A, n=$	Given A to find F $A \xrightarrow{i} F, n=$	Given A to find P $A \xrightarrow{i} P, n=$	N
1	1.0600	0.94340	1.000000	1.060000	1.0000	0.9434	1
2	1.1236	0.89000	0.485437	0.545437	2.0600	1.8334	2
3	1.1910	0.83962	0.314110	0.374110	3.1836	2.6730	3
4	1.2625	0.79209	0.228591	0.288591	4.3746	3.4651	4
5	1.3382	0.74726	0.177396	0.237396	5.6371	4.2124	5
6	1.4185	0.70496	0.143363	0.203363	6.9753	4.9173	6
7	1.5036	0.66506	0.119135	0.179135	8.3938	5.5824	7
8	1.5938	0.62741	0.101036	0.161036	9.8975	6.2098	8
9	1.6895	0.59190	0.087022	0.147022	11.4913	6.8017	9
10	1.7908	0.55839	0.075868	0.135868	13.1808	7.3601	10
11	1.8983	0.52679	0.066793	0.126793	14.9716	7.8869	11
12	2.0122	0.49697	0.059277	0.119277	16.8699	8.3838	12
13	2.1329	0.46884	0.052960	0.112960	18.8821	8.8527	13
14	2.2609	0.44230	0.047585	0.107585	21.0151	9.2950	14
15	2.3966	0.41727	0.042963	0.102963	23.2760	9.7122	15
16	2.5404	0.39365	0.038952	0.098952	25.6725	10.1059	16
17	2.6928	0.37136	0.035445	0.095445	28.2129	10.4773	17
18	2.8543	0.35034	0.032357	0.092357	30.9057	10.8276	18
19	3.0256	0.33051	0.029621	0.089621	33.7600	11.1581	19
20	3.2071	0.31180	0.027185	0.087185	36.7856	11.4699	20
21	3.3996	0.29416	0.025005	0.085005	39.9927	11.7641	21
22	3.6035	0.27751	0.023046	0.083046	43.3923	12.0416	22
23	3.8197	0.26180	0.021278	0.081278	46.9958	12.3034	23
24	4.0489	0.24698	0.019679	0.079679	50.8156	12.5504	24
25	4.2919	0.23300	0.018227	0.078227	54.8645	12.7834	25
26	4.5494	0.21981	0.016904	0.076904	59.1564	13.0032	26
27	4.8223	0.20737	0.015697	0.075697	63.7058	13.2105	27
28	5.1117	0.19563	0.014593	0.074593	68.5281	13.4062	28
29	5.4184	0.18456	0.013580	0.073580	73.6398	13.5907	29
30	5.7435	0.17411	0.012649	0.072649	79.0582	13.7648	30

TABLE 2: 8% COMPOUND INTEREST FACTORS

1	2	3	4	5	6	7	8
	Single payment		Uniform series				
	Compound amount factor	Present worth factor	Sinking fund factor	Capital recovery factor	Compound amount factor	Present worth factor	
N	Given P to find F $P \xrightarrow{i} F, n=$	Given F to find P $F \xrightarrow{i} P, n=$	Given F to find A $F \xrightarrow{i} A, n=$	Given P to find A $P \xrightarrow{i} A, n=$	Given A to find F $A \xrightarrow{i} F, n=$	Given A to find P $A \xrightarrow{i} P, n=$	N
1	1.0800	0.92593	1.000000	1.080000	1.0000	0.9259	1
2	1.1664	0.85734	0.480769	0.560769	2.0800	1.7833	2
3	1.2597	0.79383	0.308034	0.388034	3.2464	2.5771	3
4	1.3605	0.73503	0.221921	0.301921	4.5061	3.3121	4
5	1.4693	0.68058	0.170456	0.250456	5.8666	3.9927	5
6	1.5869	0.63017	0.136315	0.216315	7.3359	4.6229	6
7	1.7138	0.58349	0.112072	0.192072	8.9228	5.2064	7
8	1.8509	0.54027	0.094015	0.174015	10.6366	5.7466	8
9	1.9990	0.50025	0.080080	0.160080	12.4876	6.2469	9
10	2.1589	0.46319	0.069029	0.149029	14.4866	6.7101	10
11	2.3316	0.42888	0.060076	0.140076	16.6455	7.1390	11
12	2.5182	0.39711	0.052695	0.132695	18.9771	7.5361	12
13	2.7196	0.36770	0.046522	0.126522	21.4953	7.9038	13
14	2.9372	0.34046	0.041297	0.121297	24.2149	8.2442	14
15	3.1722	0.31524	0.036830	0.116830	27.1521	8.5595	15
16	3.4259	0.29189	0.032977	0.112977	30.3243	8.8514	16
17	3.7000	0.27027	0.029629	0.109629	33.7502	9.1216	17
18	3.9960	0.25025	0.026702	0.106702	37.4502	9.3719	18
19	4.3157	0.23171	0.024128	0.104128	41.4463	9.6036	19
20	4.6610	0.21455	0.021852	0.101852	45.7620	9.8181	20
21	5.0338	0.19866	0.019832	0.099832	50.4229	10.0168	21
22	5.4365	0.18394	0.018032	0.098032	55.4568	10.2007	22
23	5.8715	0.17032	0.016422	0.096422	60.8933	10.3711	23
24	6.3412	0.15770	0.014978	0.094978	66.7648	10.5288	24
25	6.8485	0.14602	0.013679	0.093679	73.1059	10.6748	25
26	7.3964	0.13520	0.012507	0.092507	79.9544	10.8100	26
27	7.9881	0.12519	0.011448	0.091448	87.3508	10.9352	27
28	8.6271	0.11591	0.010489	0.090489	95.3388	11.0511	28
29	9.3173	0.10733	0.009619	0.089619	103.9659	11.1584	29
30	10.0627	0.09938	0.008827	0.088827	113.2832	11.2578	30

TABLE 3: 10% COMPOUND INTEREST FACTORS

1	2	3	4	5	6	7	8
	Single payment		Uniform series				
	Compound amount factor	Present worth factor	Sinking fund factor	Capital recovery factor	Compound amount factor	Present worth factor	
N	Given P to find F $P \xrightarrow{i} F,n=$	Given F to find P $F \xrightarrow{i} P,n=$	Given F to find A $F \xrightarrow{i} A,n=$	Given P to find A $P \xrightarrow{i} A,n=$	Given A to find F $A \xrightarrow{i} F,n=$	Given A to find P $A \xrightarrow{i} P,n=$	N
1	1.1000	0.90909	1.000000	1.100000	1.0000	0.9091	1
2	1.2100	0.82645	0.476190	0.576190	2.1000	1.7355	2
3	1.3310	0.75131	0.302115	0.402115	3.3100	2.4869	3
4	1.4641	0.68301	0.215471	0.315471	4.6410	3.1699	4
5	1.6105	0.62092	0.163797	0.263797	6.1051	3.7908	5
6	1.7716	0.56447	0.129607	0.229607	7.7156	4.3553	6
7	1.9487	0.51316	0.105405	0.205405	9.4872	4.8684	7
8	2.1436	0.46651	0.087444	0.187444	11.4359	5.3349	8
9	2.3579	0.42410	0.073641	0.173641	13.5795	5.7590	9
10	2.5937	0.38554	0.062745	0.162745	15.9374	6.1446	10
11	2.8531	0.35049	0.053963	0.153963	18.5312	6.4951	11
12	3.1384	0.31863	0.046763	0.146763	21.3843	6.8137	12
13	3.4523	0.28966	0.040779	0.140779	24.5227	7.1034	13
14	3.7975	0.26333	0.035746	0.135746	27.9750	7.3667	14
15	4.1772	0.23939	0.031474	0.131474	31.7725	7.6061	15
16	4.5950	0.21763	0.027817	0.127817	35.9497	7.8237	16
17	5.0545	0.19784	0.024664	0.124664	40.5447	8.0216	17
18	5.5599	0.17986	0.021930	0.121930	45.5992	8.2014	18
19	6.1159	0.16351	0.019547	0.119547	51.1591	8.3649	19
20	6.7275	0.14864	0.017460	0.117460	57.2750	8.5136	20
21	7.4002	0.13513	0.015624	0.115624	64.0025	8.6487	21
22	8.1403	0.12285	0.014005	0.114005	71.4027	8.7715	22
23	8.9543	0.11168	0.012572	0.112572	79.5430	8.8832	23
24	9.8497	0.10153	0.011300	0.111300	88.4973	8.9847	24
25	10.8347	0.09230	0.010168	0.110168	98.3471	9.0770	25
26	11.9182	0.08391	0.009159	0.109159	109.1818	9.1609	26
27	13.1100	0.07628	0.008258	0.108258	121.0999	9.2372	27
28	14.4210	0.06934	0.007451	0.107451	134.2099	9.3066	28
29	15.8631	0.06304	0.006728	0.106728	148.6309	9.3696	29
30	17.4494	0.05731	0.006079	0.106079	164.4940	9.4269	30

TABLE 4: 12% COMPOUND INTEREST FACTORS

1	2	3	4	5	6	7	8
	Single payment		Uniform series				
	Compound amount factor	Present worth factor	Sinking fund factor	Capital recovery factor	Compound amount factor	Present worth factor	
N	Given P to find F $P \xrightarrow{i} F,n=$	Given F to find P $F \xrightarrow{i} P,n=$	Given F to find A $F \xrightarrow{i} A,n=$	Given P to find A $P \xrightarrow{i} A,n=$	Given A to find F $A \xrightarrow{i} F.n=$	Given A to find P $A \xrightarrow{i} P,n=$	N
1	1.1200	0.89286	1.000000	1.120000	1.0000	0.8929	1
2	1.2544	0.79719	0.471698	0.591698	2.1200	1.6901	2
3	1.4049	0.71178	0.296349	0.416349	3.3744	2.4018	3
4	1.5735	0.63552	0.209234	0.329234	4.7793	3.0373	4
5	1.7623	0.56743	0.157410	0.277410	6.3528	3.6048	5
6	1.9738	0.50663	0.123226	0.243226	8.1152	4.1114	6
7	2.2107	0.45235	0.099118	0.219118	10.0890	4.5638	7
8	2.4760	0.40388	0.081303	0.201303	12.2997	4.9676	8
9	2.7731	0.36061	0.067679	0.187679	14.7757	5.3282	9
10	3.1058	0.32197	0.056984	0.176984	17.5487	5.6502	10
11	3.4785	0.28748	0.048415	0.168415	20.6546	5.9377	11
12	3.8960	0.25668	0.041437	0.161437	24.1331	6.1944	12
13	4.3635	0.22917	0.035677	0.155677	28.0291	6.4235	13
14	4.8871	0.20462	0.030871	0.150871	32.3926	6.6282	14
15	5.4736	0.18270	0.026824	0.146824	37.2797	6.8109	15
16	6.1304	0.16312	0.023390	0.143390	42.7533	6.9740	16
17	6.8660	0.14564	0.020457	0.140457	48.8837	7.1196	17
18	7.6900	0.13004	0.017937	0.137937	55.7497	7.2497	18
19	8.6128	0.11611	0.015763	0.135763	63.4397	7.3658	19
20	9.6463	0.10367	0.013879	0.133879	72.0524	7.4694	20
21	10.8038	0.09256	0.012240	0.132240	81.6987	7.5620	21
22	12.1003	0.08264	0.010811	0.130811	92.5026	7.6446	22
23	13.5523	0.07379	0.009560	0.129560	104.6029	7.7184	23
24	15.1786	0.06588	0.008463	0.128463	118.1552	7.7843	24
25	17.0001	0.05882	0.007500	0.127500	133.3339	7.8431	25
26	19.0401	0.05252	0.006652	0.126652	150.3339	7.8957	26
27	21.3249	0.04689	0.005904	0.125904	169.3740	7.9426	27
28	23.8839	0.04187	0.005244	0.125244	190.6989	7.9844	28
29	26.7499	0.03738	0.004660	0.124660	214.5828	8.0218	29
30	29.9599	0.03338	0.004144	0.124144	241.3327	8.0552	30

To use the **Description of Compound Interest Factors** it is necessary to refer to tables of compound interest such as those in WALKERS PRACTICAL ACCOUNTING. Similar tables may be found in a variety of sources. As a matter of convenience for solving the quizzes in this chapter, Tables 1 through 4 give the required reference data. Column one of these tables is "N" or the number of years applicable, Column 2 is **P to find F,** Column 3 is **F to find P,** Column 4 is **F to find A,** Column 5 is **P to find A,** Column 6 is **A to find F,** and Column 7 is **A to find P.** The last column 8 is the same as 1. It is recommended that the exam candidate become thoroughly familiar with these operations by working through these quizzes several times.

RECONCILING THE BANK STATEMENT

Reconciling the bank statement is a basic function in any bookkeeping system. Some questions related to these procedures will usually appear on the contractors examination. A good discussion on this subject may be found in the aforementioned BUILDER'S GUIDE TO ACCOUNTING by Thomsett in Chapters 9 and 18. Here is a typical problem related to reconciling the bank statement:

Continental Construction Company does its banking with City Commercial Bank. On January 5 1988 Continental received its December 1987 bank statement showing a balance of $13,701. An incorrect charge of $200 was made by City Bank against Continental's account. On December 29 1987 the bookkeeper received a check in the amount of $450 but left it in a drawer until January 4 when it was deposited into the Continental account. As of December 31 1987, outstanding checks totalled $2,308. Deposits not posted by December 31 amounted to $3,175. There was a "service charge" of $15 for the last month and a $37 item marked "uncollected funds". The balance shown in Continental Construction Company's checkbook on December 31 1987 is $14,820. What is the actual balance as of 12-31-88?

(A) Unable be computed without additional information
(B) $14,560
(C) $14,768
(D) $14,820.

For the answer to this problem see Table 5, following.

COST ESTIMATING

Usually, when cost estimating questions appear on examinations, they will be part of a larger problem related to plan reading or job design. Such questions are meant to test the candidate's general understanding of how to put a job together in a business

TABLE 5

ANSWER TO BANK RECONCILIATION PROBLEM

CHECKBOOK BALANCE			BALANCE PER BANK			
Ending Balance per Books	$14,820		Ending Bank Statement Balance (last figure on bank statement)			$13,701
Add:			Add:			
Recording of errors that understate balance	- — -		Deposits in transit (amount not yet credited by bank)		3,175	
Proceeds of notes collected by bank or other items credited (added) by bank but not yet updated in checkbook	- — -	- — -	Bank errors		200	3,375
Deduct:			Deduct:			
Recording of errors that overstate balance	- — -		List of outstanding checks (amount not yet debited by bank)		2,308	
Service charges	15		Bank errors		- — -	
Printing charges	- — -					- 2,308
NSF, check, etc., or other items debited (charged) by bank but not yet updated in checkbook	37	- 52				
Reconciled Balance (Adjusted Balance)	$14,768		Reconciled Balance (Adjusted Balance)			$14,768

TABLE 6

PROFIT - PERCENTAGE CONVERSION

% MARKUP ON COST	% PROFIT ON SELLING PRICE	% MARKUP ON COST	% PROFIT ON SELLING PRICE
5.00	4.75	31.58	24.00
7.50	7.00	33.33	25.00
10.00	9.00	35.00	26.00
11.11	10.00	37.50	27.25
12.36	11.00	40.00	28.50
12.50	11.12	42.86	30.00
13.63	12.00	45.00	31.00
14.95	13.00	47.00	32.00
16.28	14.00	50.00	33.33
16.43	14.25	52.85	35.00
17.65	15.00	55.00	35.50
19.05	16.00	60.00	37.50
20.00	16.67	65.00	39.50
20.49	17.00	66.66	40.00
21.96	18.00	70.00	41.00
23.46	19.00	75.00	42.75
25.00	20.00	80.00	44.50
26.58	21.00	85.00	46.00
28.21	22.00	90.00	47.50
29.88	23.00	100.00	50.00

like manner, and are not concerned with detailed costing for competitive bidding. Figure 4 illustrates an estimating form.

After totalling all labor, material, subcontracts, and other direct job costs on a suitable estimating form, the desired markup is then added to these costs to find the correct selling price. Markup is a key work in any question or problem on cost estimating. The markup is the total amount of overhead margin and profit margin; it is a different figure for each individual business. To find the markup, first determine what the percent of overhead is. The percent of overhead may only be known from your accountant's Statement of Condition. The percent overhead added to the desired percent profit (net profit) equals the total percent markup.

EXAMPLE: Assume a mechanical construction cost totals $30.000.00, the percentage of overhead from the statement is 18%, the desired net profit is 10%. Find the correct selling price based on costs.

SOLUTION:

$$\frac{\$30,000}{1} \times \frac{28}{100} = \$8,400$$

$30,000 + $8,400 = $38,400, the selling price or, $30,000 x 1.28 = $38,400

In the above example the $8,400 markup is equal to 28% profit on cost price. However, if the 28% profit is to be realized on the selling price, then; the markup--and consequently the selling price--would be much higher.

EXAMPLE: Using the figures in the above example, find the correct selling price based on 28% profit on the selling price.

SOLUTION: Let 100% = selling price
28% = margin
72% = cost price or $30,000

then;

$$100\% = \frac{\$30,000}{.72} = \$41,666, \text{ the selling price,}$$

leaving a markup of $11,666 or $3,266 more than if the profit was based on a percentage of cost.

Table 6 gives the conversions from % markup on cost to % markup on the selling price. Correct estimating and job costing procedures require much more detail analysis than that covered in the present discussion. Actually, the overhead should be broken out and assigned in separate quantities against labor and material. Table 7 shows precalculated multipliers for finding the selling price where the overhead percentage is known.

TABLE 7

PRECALCULATED MULTIPLIERS TO FIND SELLING PRICE

PERCENT OVERHEAD FROM STATEMENT	NET PROFIT REQUIRED BEFORE TAXES								
	4%	5%	6%	7%	8%	9%	10%	12%	15%
10	1.16	1.18	1.19	1.20	1.22	1.23	1.25	1.28	1.33
11	1.18	1.19	1.20	1.22	1.23	1.25	1.27	1.30	1.35
12	1.19	1.20	1.22	1.23	1.25	1.27	1.28	1.32	1.37
13	1.20	1.22	1.23	1.25	1.27	1.28	1.30	1.33	1.39
14	1.22	1.23	1.25	1.27	1.28	1.30	1.32	1.35	1.41
15	1.23	1.25	1.27	1.28	1.30	1.32	1.33	1.37	1.43
16	1.25	1.27	1.28	1.30	1.32	1.33	1.35	1.39	1.45
17	1.27	1.28	1.30	1.32	1.33	1.35	1.37	1.41	1.47
18	1.28	1.30	1.32	1.33	1.35	1.37	1.39	1.43	1.49
19	1.30	1.32	1.33	1.35	1.37	1.39	1.41	1.45	1.52
20	1.32	1.33	1.35	1.37	1.39	1.41	1.43	1.47	1.54
21	1.33	1.35	1.37	1.39	1.41	1.43	1.45	1.49	1.56
22	1.35	1.37	1.39	1.41	1.43	1.45	1.47	1.52	1.59
23	1.37	1.39	1.41	1.43	1.45	1.47	1.49	1.54	1.61
24	1.39	1.41	1.43	1.45	1.47	1.49	1.52	1.56	1.64
25	1.41	1.43	1.45	1.47	1.49	1.52	1.54	1.59	1.67
26	1.43	1.45	1.47	1.49	1.52	1.54	1.56	1.61	1.70
27	1.45	1.47	1.49	1.52	1.54	1.56	1.59	1.64	1.72
28	1.47	1.49	1.52	1.54	1.56	1.59	1.61	1.67	1.75
29	1.49	1.52	1.54	1.56	1.59	1.61	1.64	1.70	1.79
30	1.52	1.54	1.56	1.59	1.61	1.64	1.67	1.72	1.82

Example: Total job cost is $25,000, overhead from statement is 15%, net profit required for this job is 6%. Find profit and selling price.

Solution: From the table find the multiplier = 1.27, then; $25,000 x 1.27 = $31,750 selling price. The markup is $31,750 - 25,000 = $6,750 profit.

FORM E66

ESTIMATE   PRICING MATERIAL & LABOR

PROJECT _____

ADDRESS _____

_____

DATE IN _____ DATE DUE _____ ESTIMATOR _____

JOB NO. _____

SHEET _____ OF _____ SHEETS

_____

_____

CHECKED BY _____

QUAN	DESCRIPTION	LABOR		MATERIAL		EXTENSION
		UNIT	TOTAL	UNIT	TOTAL	
	MISCELLANEOUS MATERIAL AND LABOR					
	NON PRODUCTIVE LABOR					
	JOB OVERHEAD					
	SUB TOTAL: COST OF JOB					
	% GENERAL OVERHEAD					
	MATERIAL/LABOR RATIO					
	% PROFIT ON LABOR					
	% PROFIT ON MATERIAL					
	SUB CONTRACTS					
	% PROFIT ON SUBS					
	SALES TAX					
	SERVICE RESERVE					
	GRAND TOTAL: SELLING PRICE					

FIGURE 4

ESTIMATING QUIZ NO. 1

1. GIVEN:

Subcontractors total bid	$7,950.00
Subcontract performance bond rate	0.75%
Labor estimate	$6,840.00
Labor, taxes and insurance rate	14%
Materials estimate	$24,300.00
Materials sales tax rate	5%
Total general and administrative cost estimate	$2,500.00
Overhead and contingencies	$2,800.00
Profit markup	10%
Bond on total job (first $20,000.00)	1%
Bond on total job (over $20,000.00) rate	.75%

If all costs to be considered are included in the figures above, what is the total bid on the project? Select the closest answer.

(A) $51,449.78   (C) $ 89,480.01
(B) $51,719.11   (D) $100,843.04

2. Using the principles of BUILDER'S GUIDE TO ACCOUNTING, and assuming only deductions for Social Security and Federal Income Tax, what would be the total amount of the <u>net</u> payment to the five employees shown below? Use the marital status, number of exemptions claimed, hourly wage rate, regular hours, and overtime hours as listed below. Assume no one has exceeded the limit for Social Security. Calculate overtime as 1½ times regular rate. Payments are semimonthly. Use the wage bracket tables in Circular E, January 1988 for calculating Federal Income Tax. Select the closest answer.

NAME	MARITAL STATUS	TOTAL NO OF EXEMPTIONS CLAIMED	HOURLY WAGE RATE	REGULAR HOURS	OVERTIME HOURS
Fred Jones	Married	2	$10.50	80	8
Tina William	Single	1	$ 7.25	80	5
Bill Smith	Married	3	$11.75	53	3
Pam Johnson	Married	5	$ 6.45	65	9
Jim Henry	Single	1	$12.28	80	0

(A) $3024.62   (C) $3046.08
(B) $3043.77   (D) $3052.41

ESTIMATING QUIZ NO. 2

1. GIVEN:

General contractor, office overhead	$ 65,000.00
Job conditions and job overhead	125,000.00
Plant tools and equipment	90,000.00
Subcontractors total bid	750,000.00
Subcontractors performance bond rate	0.60%
Labor estimate	450,000.00
Labor insurance and tax rate	21%
Material estimate	1,025,000.00
Material tax rate	5.0%
Profit markup	3.0%

Performance Bond:

1st $100,000.00 @ $9.25 per thousand
2nd $400,000.00 @ $7.50 per thousand
3rd $2,000,000.00 @ $7.00 per thousand
4th over $2,500,000.00 @ $6.00 per thousand

Select the closest answer:

(A)   less than $2,500,000.00

(B) $2,657,326.75

(C) $2,709,245.45

(D) $2,754.241.95

(E) $2,799,241.95

ESTIMATING QUIZ NO. 3

1. GIVEN: A General Contractor requests bids on the
    following sitework:

> Concrete walks
> Fencing
> Sodding and Seeding
> Landscaping

Subcontractor "A" submits a bid on the concrete walks
and fencing (only) for a total of $6,500.00.

Subcontractor "B" submits a bid on the fencing and
landscaping (only) for a toal of $4,000.00.

Subcontractor "C" submits a bid on the concrete
walks, sodding, and seeding (only) for a total of
$6,000.00.

Subcontractor "D" submits a bid on the sodding,
seeding, and landscaping for a total of $3,500.00.

Subcontractor "E" submits a bid on the fencing,
sodding, seeding, and landscaping (only) for a
total of $5,500.00.

The General Contractor estimates that his cost to
provide the concrete walks himself to be $4,400.00.

The General Contractor estimates that his cost to
provide the fencing himself to be $1,900.00.

The General Contractor estimates that his cost to
provide the sodding and seeding himself to be
$1,600.00.

The General Contractor estimates that his cost to
provide the landscaping himself to be $2,000.00.

If the bids and estimates given above are the only costs to
be considered, the lowest combination of bids and/or
estimates to provide all of the items listed is _____.

> (A) $ 9,600.00
> (B) $ 9,700.00
> (C) $ 9,800.00
> (D) $ 9,900.00
> (E) $10,000.00

ACCOUNTING QUIZ NO. 1

1. According to BUILDER'S GUIDE TO ACCOUNTING arrange the following accounts into Assets, current and fixed, and Liabilities, current and long term.

          Accounts payable   Cash in banks
          Notes payable   Accounts receivable
          Prepaid insurance   Deposits on account
          Accrued taxes   Secured loan
          Office equipment   Unsecured loan
          Securities   Inventory
          Reserve for depreciation   Land and building
          Mortgage loan   Machinery

2. The ABC Construction Co. at the close of the second quarter 1984, has the following balances. If the balances represent net amounts after depreciation, what is the net worth of company for June 1984?

          Land and building $112,000.00
          Mortgage loan   42,000.00
          Machinery   7,000.00
          Cash   4,500.00
          Taxes payable   1,450.00
          Accounts receivable   3,700.00
          Accounts payable   1,100.00
          Prepaid insurance   750.00

3. ABC Construction Co. decides to invest in additional machinery costing $11,500.00, financing 70%. Using the figures in problem 2 above, what is the current net worth after the purchase.

4. If you loaned out $1800 at 8% , how many years would it take for the interest to double the principle?

   A. 6 years   B. 7 years   C. 8 years.   D. 9 years

5. The following accounts are an example of accounts and are found on the

          Office salaries
          Depreciation
          Rent
          Utilities
          Payroll taxes
          Equipment maintenance

   A. Asset/Balance Sheet   D. Expense/Income Statement
   B. Income/Income Statement   E. Capital/Balance Sheet
   C. Liability/Balance Sheet

ACCOUNTING QUIZ NO. 2

If you deposit a single payment of $12,650 in a fund at 6%
per annum, what would the money be worth at the end of 18
years?

    (A) $34,729.40   (C) $48,573.12
    (B) $36,106.90   (D) $51,690.88

2. If you deposit $2000 per year into a fund at 10%, at the
  end of 20 years how much money you would have?

    (A) $108,976.00   (C) $114,550.00
    (B) $112,862.10   (D) $116,550.20

3. If you can afford to pay $450 per month and you wish to
  make a 5 year loan at 8.% how much money (net) would you
  receive.

    (A) $30,375.82   (C) $21,560.58
    (B) $28,660.43   (D) $23,500.00

4. In problem 3 how much would you pay for the use of the
  money?

    (A) $6,332.67   (C) $5,972.63
    (B) $5,439.42   (D) $4,522.40

5. If you were to borrow $60,000 for 30 years at 10% your
  monthly payment would be?

    (A) $753.83   (C) $533.95
    (B) $694.40   (D) $560.62

6. How much more would your monthly payment be in problem
  5 if you borrowed the money for 15 years at the same rate
  of interest?

    (A) $123.42   (C) $643.80
    (B) $354.62   (D) $753.83

7. If $13,000 is needed to pay off a Balloon Note 5 years
  from now, how much money at 12% interest must be set aside
  at the end of each year in order to make the payment?

    (A) $7,134.40   (C) $3,673,57
    (B) $2,600.00   (D) $2,046.33

ACCOUNTING QUIZ NO. 3

1. If a contractor secured a loan for $12,000 at 16.5%
   simple interest payable at the end of 90 days, his total
   payment would be_____.

A.	$12,848.16	C.	$13,448.18
B.	$13,980.00	D.	$12,488.16

2. GIVEN: A contracting firm has the following account
   balances:

Accounts payable	$30,700.00
Loans receivable	6,000.00
Payroll taxes payable	1,500.00
Prepaid insurance	400.00
Cash in bank	14,300.00
Inventory - Materials	10,000.00
Accrued FICA taxes	2,500.00
Prepaid taxes	3,000.00
Autos and trucks	25,000.00
Land held for development	50,000.00
Notes payable	15,000.00

   If the above balances are the only amounts to be
   considered, the firm's net worth is _____.

A.	$41,000	D.	$53,000
B.	$59,000	E.	$61,000

Note: Questions 3, 4, and 5 refer to the Yearly Statement of
Accounts for 1981 and Yearly Income Analysis for 1981 on the
following page.

3. GIVEN: A contractor had a net worth of $25,665.40 for
   1980. The contractor tabulates his assets and
   liabilities for 1981 and discovers his net worth
   for 1981 represents a _____his net worth for
   1980.

   A. loss of $2,347.55 from
   B. gain of $2,347.55 over
   C. gain of $3,427.49 over
   D. loss of $3,427.49 from
   E. gain of $5,460.55 over

4. A contractor determined his net income for the year and
   realized a gross profit for his construction contracts
   (only) of_____.

A.	$21,506.88	C.	$20,306.58
B.	$18,606.38	D.	$43,235.22

5. In 1981 a contractor borrows $1,000 and pays $150.00 interest charges on the loan. In the same year he receives $566.00 as interest due him. In preparing his Yearly Income Analysis for 1981, he figures his total general expenses to be _____ .

    A. $9,348.51   C. $10,998.41
    B. $9,764.51   D. $8,198.41

YEARLY STATEMENT OF ACCOUNTS FOR 1981

ASSETS		LIABILITIES	
Current Assets		**Current Liabilities**	
Cash in bank	$18,776.51	Accounts payable	$3,464.32
Accounts receivable	3,650.02	Notes Payable	1,000.00
Inventory - materials	10,433.00	Payroll taxes	575.29
Due from contracts	4,888.82	Property taxes	2,320.04
Fixed Assets		**Long-Term Liabilities**	
Furniture and Fixture	376.34	Mortgage payable	34,650.00
Trucks	1,898.20		
Building	10,550.00	Total Liabilities	42,009.65
Mortgage	12,000.00		
Total Assets	$65,327.50	Net Worth	

YEARLY INCOME ANALYSIS FOR 1981

Construction Contract Income		$37,450.90
Cost of Contracts:		
Materials	12,624.30	
Sub-contractors	4,520.02	
Gross Profit		
General Expenses		
Rent	5,237.00	
Utilities	1,020.06	
Truck expense	2,151.36	
Truck depreciation	790.09	
Interest paid on borrowed money		
Total General Expenses		
Other Income:		
Purchase discounts	1,500.00	
Interest earned		
Total Yearly Income		$13,024.07

ACCOUNTING QUIZ NO. 4

Note: Questions 1 thru 4 refer to the attached Budget Statement.

1. The total excess cash (or deficiency) available for expenses over total cash needed before financing in the first quarter would be_____.

    A. An excess of $7,650.00
    B. A deficiency of $7,650.00
    C. An excess of $8,450.00
    D. A deficiency of $8,450.00

2. The cash balance at the end of the fourth quarter is _____.

    A. $8,800.00
    B. $9,500.00
    C. $9,762.50
    D. $9,934.00

3. The net effect of financing for the fiscal year is _____.

    A. An excess of $28,000.00
    B. An excess of $12,800.00
    C. A deficiency of $12,900.00
    D. A deficiency of $12,800.00

4. The total cash needed was what percent of the cash available for expenses in the second quarter?

    A. 72.4%
    B. 80.6%
    C. 85.5%
    D. 88.6%
    E. 95.4%

YEARLY BUDGET OF INCOME AND EXPENSES FOR XYZ COMPANY

FISCAL YEAR 1982

	Quarters			
	1	2	3	4
Beginning Cash Balance	8,500.00	3,550.00	7,260.00	5,600.00
Income from completed contracts.	181,500.00	232,010.00	240,000.00	146,000.00
Total	190,000.00	235,500.00	247,260.00	151,600.00
Cash retained by contractor from cash receivables.	10,000.00	15,000.00	25,000.00	6,000.00
Cash abailable for expenses	180,000.00	220,560.00	222,260.00	145,600.00
Expenses				
Materials	43,000.00	48,500.00	40,000.00	18,100.00
Labor	54,500.00	46,000.00	65,610.00	42,000.00
Job overhead	33,150.00	49,000.00	53,650.00	34,500.00
Fixed Overhead	52,000.00	52,000.00	52,000.00	52,000.00
Total	182,650.00	195,500.00	211,260.00	146.600.00
Minimum Cash balance desired	5,000.00	5,000.00	5,000.00	5,000.00
Total Cash Needed	187,650.00	200,500.00	216,260.00	151,600.00
Total excess cash available for expenses over total cash needed before financing (Deficiency)		20,060.00	6,000.00	(6,000.00)
Financing				
Cash to be borrowed (Less repayments)	12,000.00 -0-	-0- (12,000.00)	-0- -0-	16,000.00 -0-
(Less Interest payments)	(800.00)	(800.00)	(400.00)	(1,200.00
Net effect of Financing	11,200.00	(12,800.00)	(400.00)	14,800.00
Cash Balance End of Quarter	3,550.00	7,260.00	5,600.00	

Note: All amounts in ( ) are negative amounts.

ANSWER SHEET

ESTIMATING QUIZ NO. 1

1.
Subcontractor's bid	$7,950.00
Subcontractor's bond (.0075 x 7,950)	59.63
Labor	6,840.00
Labor tax & insurance (.14 x 6,840)	957.60
Material	24,300.00
Material tax (.05 x 24,3000)	1,215.00
General overhead	2,500.00
Job overhead	2,800.00
Sub sub total	$46,622.23
Mark up (.10 x 46,622.23)	4,662.22
Sub total	$51,284.45
Bond (.01 x 20,000)	200.00
Bond [.0075 x (51,284.48 - 20,000)]	234.63
TOTAL BID	$51,719.08

ANSWER B

2.
	Hours				RATE		GROSS	
	Regular		O.T.					
Jones	80	+	12.0	= 92.0	x	$10.50	=	$966.00 minus
Williams	80	+	7.5	= 87.5	x	$ 7.25	=	$634.38 minus
Smith	53	+	4.5	= 54.5	x	$11.75	=	$675.73 minus
Johnson	65	+	13.5	= 78.5	x	$ 6.45	=	$506.33 minus
Henry	80	+	0.0	= 80.0	x	$12.25	=	$980.00 minus

	Income Tax		Social Security		Net
Jones	116.00	minus	$ 69.07	=	780.93
Williams	81.00	minus	45.36	=	508.02
Smith	56.00	minus	48.32	=	571.41
Johnson	20.00	minus	36.21	=	450.12
Henry	168.00	minus	70.07	=	741.93
			TOTAL		$3,052.41

ANSWER D

Note: The time expended on these questions should require about 10 to 12 minutes each.

ANSWER SHEET

ESTIMATING QUIZ NO. 2

1. | | |
   |---|---:|
   | General contractor, overhead | $65,000.00 |
   | Job conditions and job overhead | 125,000.00 |
   | Plant, tools, and equipment | 90,000.00 |
   | Sub contractor total bid | 750,000.00 |
   | Subs performance bond = $750.000 x .006 | 4,500.00 |
   | Labor estimate | 450,000.00 |
   | Labor insurance and tax rate = $450,000 x .21 | 94,500.00 |
   | Material estimate | 1,025,000.00 |
   | Material tax rate = $1,025,000 x .05 | 51,250.00 |

SUB TOTAL	$2,655,250.00
	79,657.50
Profit markup = $2,655,250 x .03	$2,734,907.50

   * Performance bond

100,000/1000=	100 x $9.25	$925.00	
400,000/1000=	400 x $7.50	3,000.00	
2,000,000/1000=	2,000 x $7.00	14,000.00	
234,707.50/1000=	234.91 x $6.00	1,409.45	
		$19,334.45	19,344.45

TOTAL	$2,754,241.95

   ANSWER D

* Performance bond is based on dollars per thousand;
   drop the last three zero's.

ANSWER SHEET

ESTIMATING QUIZ NO. 3

	A	B	C	D	E	GC1	GC2	GC3	GC4
CONCRETE WALKS	X		X			X			
FENCE	X	X			X		X		
SOD & SEED			X	X	X			X	
LANDSCAPE		X		X	X				X
	6500	4000	6000	3500	5500	4400	1900	1600	2000

POSSIBLE COMBINATIONS:

A + D		=	$10,000.00
A + GC3		=	$10,100.00
B + C		=	$10,000.00
B + GC1 + GC3	=	$10,000.00
C + GC2 + GC4	=	$ 9,900.00
D + GC1 + GC2	=	**$ 9,800.00***
E + GC1		=	$ 9,900.00
GC1 + GC2 + GC3 + GC4 =	$ 9,900.00

* LOWEST COMBINATION OF BIDS   ANSWER   C

ANSWER SHEET

ACCOUNTING QUIZ NO. 1

1. **CURRENT ASSETS:**

    Cash in banks
    Accounts receivable
    Securities
    Deposits
    Inventory
    Prepaid insurance

    FIXED ASSETS:

    Land and building
    Machinery
    Office equipment
    *Reserve for depreciation

    CURRENT LIABILITIES:

    Accounts payable
    Notes payable
    Accrued taxes

    LONG TERM LIABILITIES:

    Secured loans
    Unsecured loans
    Mortgage loans

2.

ASSETS		LIABILITIES & CAPITAL	
CURRENT:		**CURRENT:**	
Cash	$4,500.00	Accounts payable	$1,100.00
Accounts receivable	3,700.00	Taxes, payable	1,450.00
Prepaid insurance	750.00		
FIXED:		**LONG TERM:**	
Land and Building	112,000.00	Mortgage loan	42,000.00
Machinery	7,000.00	**TOTAL LIABS.**	44,550.00
		NET WORTH	$83,400.00
TOTAL ASSETS	$127,950.00	**TOTAL LIABS. AND CAPITAL**	$127,950.00

3. Net worth remains the same, current assets change to fixed assets.

4. Answer D, 9 years

5. Answer D

* Note that reserve for depreciation or accumulated depreciation will reduce assets by a certain amount to reflect the true value of an asset such as land, buildings, automobiles and equipment.

RESERVE OR ACCUMULATED DEPRECIATION WILL REDUCE ASSETS.

ANSWER SHEET

ACCOUNTING QUIZ NO. 2

1. From Table 1, **P to find; F:**

    $12,650 x 2.8543 = 36,106.90   ANSWER B

2. From Table 3, **A to find F:**

    $2000 x 57.2750 = 114,550.00   ANSWER C

3. From Table 2, **A to find P:**

    $450 x 12 months = $5400 per year
    $5400 x 3.9927 = 21,560.58   ANSWER C

4. Total payment $450 x 12 x 5 = $27,000.00

    $27,000.00
    $\underline{-21,560.58}$
    $ 5,439.42   ANSWER B

5. From Table 3, **P to find A:**

    $60,000 x 0.10679/12 = $533.95   ANSWER C

6. From Table 3, **P to find A:**

    $60,000 x 0.131474/12 = $657.37
    $657.37
    $\underline{-533.42}$
    $123.42   ANSWER A

7. From Table 4, **F to find A:**

    $13,000 x 0.157410 = $2046.33   ANSWER D

ANSWER SHEET

ACCOUNTING QUIZ NO. 3

1. Total payment = 12,000 x .165 = 1980
   1980/365 days x 90 days = $488.22 per
   year (interest)
   $12,000.00 + $488.22 = $12,488.22

                                        ANSWER D

2.   **ASSETS**   **LIABILITIES**

   Cash in bank $14,300.00   Accounts pay. $30,700.00
   Loans receivable   6,000.00   Payroll taxes   1,500.00
   Prepaid insurance   400.00   Accrued FICA
   Prepaid taxes   3,000.00   taxes   2,500.00
   Inventory - mats. 10,000.00   Notes payable   15,000.00
   Autos and trucks   25,000.00
   Land held
   for development   50,000.00

        NET WORTH = ASSETS - LIABILITIES

        Total Assets $108,700.00
        Total Liabilities   49,700.00
        Net Worth   59,000.00   ANSWER B

3. Net worth 1980 = $25,665.40
   Net worth 1981 = $23,317.85
                    $ 2,347.55   ANSWER A

4. Yearly Income Analysis for 1981

        Construction contract income $37,450.90
        Less materials   12,624.30
        Less sub-contractors   4,520.02
                                      $20,306.58   ANSWER C

5. $150.00 interest charge is an expense
   $566.00 interest due and paid is other income

   Total General Expenses = $9,348.51   ANSWER A

ANSWER SHEET

ACCOUNTING QUIZ NO. 4

1. B.   Cash needed $187,650.00
         Cash available   180,000.00
                               (7,650.00)

2. A.   Excess cash   (6,000.00)
         Cash borrowed   16,000.00
         Interest payment   (1,200.00)
                                8,800.00

3. B.   11,200.00
                              (12,800.00)
                                 (400.00)
                               14,800.00
                               12,800.00

4. D.   Total needed   195,500.00 = 88.6
         Cash available   220,560.00

CODES, STANDARDS AND LAWS

CODES

Code questions are, with few exceptions, taken from the local prevailing code, and may constitue a heavy proportion of questions on the exam. Different regions of the country adopt different codes and elect different code bodies. You must be completely familiar with the particular code used in the area where you take the examination.

Two kinds of laws govern the erection of buildings: immutable natural laws from which we derive our scientific and technical knowledge, and the constantly changing body of civil laws, codes, and regulations, whose chief purpose is usually safety, but in recent years include energy conservation. Code compliance may be voluntary, or when adapted by regional governments, may become law.

There are three major building-code bodies in the United States. These are Building Officials and Code Administration (BOCA), International Conference of Building Officials (ICBO), and Southern Building Code Congress International (SBCCI). Jointly, these three bodies sponsor both the Council of American Building Officials (CABO) One-and Two-Family Dwelling Code and the CABO Model Energy Code. Initially funded by the U.S. Department of Energy, the CABO codes are now maintained by CABO.

BOCA: BOCA Codes are recognized and adopted by Kansas, Oklahoma, Missouri, Illinois, Wisconsin, Michigan, Indiana, Ohio, Kentucky, Virginia, West Virginia, Vermont, New Hampshire, Connecticut, Rhode Island, and parts of Pennsylvania. BOCA issues compliance reports that are recognized in BOCA jurisdictions to show conformance with appropriate standards and BOCA codes.

SBCCI: SBCCI codes are recognized and adopted in the southeastern part of the United States; Arkansas, Louisiana, Mississippi, Alabama, Tennessee, Georgia, Florida (above the Broward County line), North Carolina, South Carolina, and smaller towns in the state of Texas. They also offer the service of compliance reports to show compliance with apporpriate standards and the "Standard" code.

ICBO: ICBO is the publisher of the Uniform Building Code and cosponsor of the Uniform Mechanical Code which is recognized and adapted by California, Oregon, Washington, Nevada, Idaho, Montana, Wyoming, North Dakota, South Dakota, Colorado, Nebraska, New Mexico, Arizona, Utah, Minnesota, Maine, and the major cities in the state of Texas.

New York and Wisconsin have their own state codes. Four Florida counties; Dade, Broward, Monroe and Collier use the South Florida Building Code (SFBC) while the rest of Florida is ruled by the SBCCI. For an overview of Code geography see the Figures following.

The candidate for any examination <u>must</u> be thoroughly familiar with the codes covering his or her local area and possess a current, updated copy of the prevailing code.

STANDARDS
Other industry standards covering a wide variety of tasks and design recommendations such as grading for lumber, span tables for joists, asphalt paving, burial and removal of underground tanks, safety code for derricks, etc., may appear on the list of required references for the exmination. Frequently, several of the AIA (American Institute of Architects) Contract Documents related to contractors and subcontractors may be on the list of required references.

LAWS
In addition to questions which test the candidate's knowledge of codes and standards, every exam will contain questions about law. These <u>law questions</u> may refer to national laws such as Income Tax and Social Security, OSHA: Occupational Safety and Health Standards, Fair Labor and Standards Act, Workmen's Compensation, Sales Tax and Use Laws, etc., as well as state and county laws covering the Mechanics Lien Law, energy conservation in buildings, diposal of underground tanks, antipollution and pollution control, contractor licensing, etc.

Questions pertaining to Federal Tax laws requiring the bookkeeping function of calculating witholding taxes will appear in almost every examination in the country. It is necessary, therefore that the examination candidate be thoroughly familiar with the document known as the **Employer Tax Guide, Circular E,** published by the Internal Revenue Service, and all the tables therein, as well as the <u>curent</u> tax rates for payroll deductions for both Social Security (FICA) and Federal Unemployment Tax (FUTA).

The **INDEX GUIDE TO EMPLOYER'S CIRCULAR E FORMS** and **SOLVING PROBLEMS OF CIRCULAR E** as well as the State and Federal Law Quiz No.1 shown below are based on the 1996 edition of Circular E. Because the Federal Tax laws are changed frequently, it is not possible to keep current with any book dealing with this law. <u>The reader must update these pages by substituting current figures for those shown.</u> But the manipulation of the tables remains the same and the example for how to work the Circular E is valid; these manipulations are not like to change.

In keeping with the authors' stated intention of using the Florida State Exam as the <u>typical</u> case for illustrative purposes, the following pages will refer to: Florida Energy Code; Chapter 489, Florida Licensing Law,Chapter 713; Mechanics Lien Law

Answers to the State and Federal Law Quiz No.1 may be found in Circular E, Florida Unemployment Compensation Law, Florida Workers Compensation Law, Mechanics Lien Law Chapter 713, and the Southern Standard Building Code.

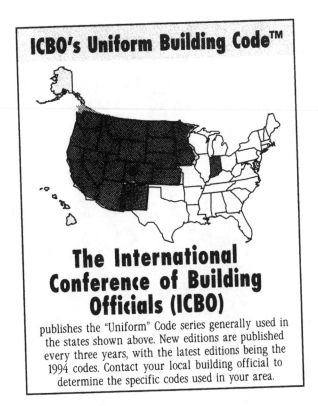

The International Conference of Building Officials (ICBO)

publishes the "Uniform" Code series generally used in the states shown above. New editions are published every three years, with the latest editions being the 1994 codes. Contact your local building official to determine the specific codes used in your area.

FIGURE 1

Building Officials and Code Administrators International (BOCA)

publishes the "National" Code series generally used in the states shown above. New editions are published every three years, with the latest editions being the 1996 codes. Contact your local building official to determine the specific codes used in your area.

FIGURE 2

The Southern Building Code Congress International (SBCCI)

publishes the "Standard " (sometimes referred to as "Southern") Code series generally used in the states shown above. New editions are published every three years, with the latest editions being the 1994 codes. Contact your local building official to determine the specific codes used in your area.

FIGURE 3

FLORIDA ENERGY EFFICIENCY CODE FOR BUILDING CONSTRUCTION
1993 Edition

The FLORIDA ENERGY EFFICIENCY CODE is designed to develop methods of conserving electrical power consumption in commercial and residential construction -- both new and remodeled buildings --
through more effiecient methods of insulation, window treatment, and the elimination of infiltration, to help reduce the cooling and heating load of the building.

The basic determination of the amount of heating and/or cooling required for a building envelope is the calculation of a heating or cooling load, wherein all the resistances of the components of the structure are calcualted using the formual Btuh = Area x U x Td (See CONCRETE MASONRY HANDBOOK, page 51). The calculations for the FLORIDA ENERGY EFFICIENCY CODE are not heating and/or cooling load calculations, but rather a point system method of staying within the limits of the required points to meet the code requirements.

Essentially, one must understand how to calculate the amount of points for each component part of the structure. The better the structure is protected from the elements of weather, the less points will accumulate. Also, if ventilating fans are used and more efficient cooling and heating equipment is used, credit points will accumulate, thereby reducing the total points.

The 1993 Edition of THE FLORIDA ENERGY EFFICIENCY CODE is composed of Chapters 1, 2, 3, 4, 5, 6, 8 and appendixes. There is no Chapter 7. According to the Preface, anyone using this Code after January 1 1994 must deleted Chapters 5 and 8. Therefore, the current user must remember that only 5 of the chapters listed are valid, these are 1 Administration and Enforcement, 2 Definitions 3. Referenced Standards, 4 Commercial Building Compliance, 6 Residential Building Compliance, 8 Simplified Annual Energy Method for Commercial Buildings.

Although the authors urge examination cadidates to highlight and tab all books, particular attention needs to be paid to tabbing and highlighting this book for several reasons: the information is poorly organized, poorly written, and does not include an index. Therefore, it is unnecessarily difficult to follow some passages. Careful highlighting and tabbing will be helpful. The candidate may also consider constructing his or her own index.

It is essential that the Credit Point Form be fully understood. It is recommended that the Residential Instruction Manual be purchased and studied to fully undestand the application of the Residential Point System.

ENERGY EFFICIENCY CODE
FOR BUILDING CONSTRUCTION 1993
QUIZ

1. Circulating hot water systems (including piping for waste heat recovery systems (HRUs) and the first_____of hot and cold water piping from the storage system shall be insulated with insulation 1" minimum thickness with a thermal conductivity no greater than 0.3 Btu/in/hr/ ft²/°F.

    A. 6'   B. 8'   C. 10'   D. 12'

2. The minimum energy factor for a 120 gallon electric water heater is_____.

    A. 0.77   B. 0.80   C. 0.43   D. 0.50

3. 1" diameter refrigerant piping with fluid below 40°F must have an insulation thickness of_____.

    A. 0.5"   B. 0.75"   C. 1"   D. 1.5"

4. Horizontal flexible duct shall be supported at intervals not greater than_____.

    A. 5'   B. 6'   C. 8'   D. 10'

5. The reinforced core shall be mechanically attached to the duct fitting by a drawband installed directly over the wire-reinforced core and the duct fitting. The duct fitting shall extend a minimum of 2" into each section of duct core. When the flexible duct is larger than 12" in diameter or the design pressure exceeds_____, The drawband shall be secured by a raised bead or indented groove on the fitting.

    A. ½" W.G.   B. 1" W.G.   C. 1.5" W.G.   D. 2" W.G.

6. Flexible duct sag between supports shall not exceed_____of length.

    A. 1/2" per 100" of length   C. 3/4 per foot of length
    B. 1/2" per foot of length   D. 1 per foot of length

7. Tapes shall be applied such that they extend not less than_____ onto each of the mated surfaces and shall totally cover the____ joint. When used on rectangular ducts, tapes shall be used only on joints between parallel rigid surfaces and on right angle joints.

    A. 1/2"   B. 1.5"   C. 1"   D. 2"

8. The minimum installed thermal resistance (R value) for air distribution system components in attics is_____.

   A. R-6   B. R-4.2   C. R-5   D. R-3

9. The minimum annual fuel utilization efficiency of a gas warm air furnace of less than 225,000 Btuh is_____.

   A. 80%   B. 81%   C. 78%   D. 75%

10. The minimum annual fuel utilization efficiency of a gas fired direct heating wall unit fan type of over 42,000 Btuh is_____

   A. 73%   B. 74%   C. 59%   D. 60%

11. A room gas-fired direct heating equipment with more than 46,000 Btuh must have a minimum AFUE of_____%.

   A. 64   B. 65   C. 73   D. 75

12. A gas fired steam boiler less than 300,000 Btuh must have a minimum AFUE of_____.

   A. 80%   B. 78%   C. 80%   D. 75%

13. Three thousand-four hundred-fourteen Btu is equivalent to_____

   A. 550 foot lbs   C. 2.13 lbs/sq ft
   B. 623 kilograms   D. 1000 watts

14. Any public building or portion thereof whose peak design rated energy usage for all purposes is less than one watt per square foot of floor area, shall be_____the FEEC.

   A. a part of   C. exempt from
   B. included in   D. in conjunction with

15. All new buildings, with enclosed ceiling assemblies shall have insulation in ceilings rated at_____or more.

   A. R-11   B. R-12   C. R-19   D. R-22

16. All exposed areas of a building envelope which enclose conditioned space, except openings for windows, skylights, doors and building service systems are known as_____.

   A. transmission gains   C. EPI
   B. reflectance   D. opaque areas

17. The allowable air infiltration rate for residential swinging doors is_____.

   A. 11.0   B. 5.60   C. 3.42   D. 0.50

18. The elements of a building which enclosed conditioned spaces through which thermal energy may be transferred to or from the exterior is known as the_____.

    A. coefficient of performance
    B. efficiency, overall system
    C. energy performance index
    D. building envelope

19. Restrooms of public facilities shall be equipped with outlet devices which limit the flow of hot water to a _____ of _____

    A. minimum   110°F
    B. maximum   0.5°F
    C. maximum   3 gpm
    D. maximum   0.5 gpm

20. The ratio of net cooling capacity in Btu/h to total rate of electric input in watts under designated operating conditions is_____.

    A. recovered energy   C. EER
    B. COP   D. AFUE

ANSWERS

ENERGY EFFICIENCY CODE 1993

1.	B	8'	¶ 612.1.ABC.5	p 6-51
2.	A	0.77 EF	Table 6-11	p 6-50
3.	C	1"thick	Table 6-10	p 6-46
4.	A	5' (flex duct)	¶ 610.1.ABC.3.3.6.(#4)	p 6-41-42
5.	B	1" W.G.	¶ 610.1.ABC.3.3.1	p 6-41
6.	B	½ " per ft	¶ 610.1.ABC.3.3.6	p 5-11
7.	C	1"	¶ 610.1.ABC.3.0.7.(#4)	p 6-39
8.	A	R-6	Table 6-9	p 6-38
9.	C	78%	Table 6-7	p 6-33
10.	B	74%	Table 6-8	p 6-34
11.	B	65%	Table 6-8	p 6-34
12.	D	75%	Table 6-7	p 6-33
13.	D		Definitions	p 2-8
14.	C		¶ 101.5.2.	p 1-5
15.	C		¶ 604.1.C.1 (#2)	p 6-14
16.	D		Definitions	p 2-10
17.	D		¶606.1.ABC.1(#3)	p 6-16
18.	D		Definitions	p 2-3
19.	D		¶412.1.ABC.2.3.2.	p 4-46
20.	C		EER	p 2-5

CHAPTER 713

PART I, MECHANICS' LIEN LAW (ss. 713.01 thru 713.37)

TIME SCHEDULE FOR POSTING, FILING OR SERVING DOCUMENTS OF THE MECHANICS' LIEN LAW

TYPE OF DOCUMENT	PERSON WHO POSTS, SERVES OR FILES	TIME TO POST, SERVE OR FILE
Notice to Owner (713.06)	Anyone entitled to a mechanics' lien except laborers	Post a notice on the owner before commencing or not later than 45 days from commencing
Claim of Lien (713.08)	Every lienor including laborers and persons in privity	File in the Clerk's* office not later than 90 days after completion of work
Notice of Delivery (713.09)	Seller and purchaser of materials delivered to a place other than the job site	Serve on owner a notice signed by both seller and purchaser
Notice of Commencement (713.135)	Owner or his authorized Agent	Record in Clerk's office before actually commencing work
Waiver of Release (713.20)	Any lienor other than a laborer; laborer may waive to the extent of labor previously performed	Any time
Notice of Contest of Lien (713.22)	Owner or his Agent or Attorney; this accelerates time for filing suit	Record in Clerk's office
Payment Bond (713.23)	Contractor	Before commencing the construction

*If the claim is for $5000 or less it must be filed with the County Clerk's office. If the claim is for more than $5000 it must be filed with the Circuit Clerk's office.

CUT OUT THIS TABLE AND PASTE INTO THE BACK OF YOUR LIEN LAW

GUIDE TO CIRCULAR E
EMPLOYER'S TAX GUIDE 1996

A thorough knowledge of Circular E is important for the license candidate. Every contractor should understand Circular E and know how to calculate payroll and tax deductions. These deductions include Federal Income Tax Withholding, Social Security, Medicare, and Unemployment taxes in conjunctio with the various state laws. Circular E, Publication 15, 1996 is available from your nearest Internal Revenue Service office or WADC-9999, Rancho Cordova, CA 95743-9999 or call 1-800-829-3676.

IMPORTANT ABBREVIATIONS TO REMEMBER

SSA	Social Security Administration
IRS	Internal Revenue Service
FRB	Federal Reserve Bank or Branch
EIC	Earned income credit
RRTA	Railroad Retirement Taxes
FUTA	Federal Unemployment TAx
FICA	Social Security Tax
FTD	Federal Tax Deposit

IMPORTANT DATES:

All employers must file Quarterly Returns of Withheld Income Tax, Social Security and Medicare Taxes, using Form 941. Due dates for returns and payments are:

QUARTERS	ENDING	DUE DATE	IF DEPOSITS ARE ON TIME RETURN MAY BE FILED ON
Jan-Feb-March	March 3	April 30	May 10
Apr-May-June	June 30	July 31	Aug 10
July-Aug-Sept	Sept 30	Oct 31	Nov 10
Oct-Nov-Dec	Dec 31	Jan 31	Feb 10

Federal Unemployment Tax, FUTA, must be filed annually by January 31, using Form 940 or 940 EZ.

WHAT THE EMPLOYER MUST WITHHOLD FROM THE EMPLOYEES::

1. Withholding Taxes. See pp 22-53, Circular E

2. Social Security, FICA;
   6.2% (.062) on the first $62,700 earnings.
   1.45% (.0145) for medicare
   7.65% (.0765) total FICA and medicare

   Each of these amounts must be matched by the employee (contributing matching funds). Therefore, the employer must make quarterly deposits of two times 7.65%, or15.3% of each employee's earned wages.

3. Federal Unemployment Tax, FUTA is not deducted from the employee but is paid wholly by the employer. The FUTA tax rate is 6.2% (.062). If the employer pays FUTA to the State, 5.4% can be credited, leaving 0.8% (.008) to be paid on the Form 940 on the first $7,000 annual wage.

SOLVING PROBLEMS OF CIRCULAR E EMPLOYER'S TAX GUIDE

When solving problems of Circular E, remember to update the figures shown here to conform with your current tax laws for the year in which you are working. The following example is taken from the 1996 Circular E.

EXAMPLE:

Assume a married employee receives a semimonthly salary of $945.00 and he claims 3 exemptions.

    A. What is the employee's net take-home home pay for the period?

    B. What is the employee's social security/medicare deduction for the year?

    C. What is the amount of social security and income tax withholding the employer must deposit for this person in one year?

SOLUTION:

    A. 1. Calculate the Social Security/Medicare tax: Rate of pay x 7.65% = $945.00 x .0765 = $72.29

        2. Figure the income tax from the Married Persons-Semimonthly Payroll Period Tables, column marked 3 exemptions, find $55.00 amount of withholding.

        3. Net take home pay = $945.00 - $72.29 - $55.00 = $817.71

    B. 1. Maximum wages subject to Social Security = $62,700

        2. Yearly earnings = $945.00 x 2 x 12 mos. = $22,680

        3. Yearly earnings are less than maximum; therefore, the yearly deduction = $22,680 x 7.65% = $1,735.02

    C. 1. Social Security tax deduction from employee equals: $1,735.02
                       Matching funds by employer equals: $1,735.02
                           Total FICA deposit equals: $3,470.04

        2. Income tax deduction = (withholding/pay period) x (pay periods/year)
           =(Income tax deposit) $55.00 x 24 = $1320.00

        3. Social Security tax $3,470.04 + Income Tax $1320.00 = $4789.92 total deposit.

           NOTE: Payroll tables are express as follows
                   Weekly payroll period   = 52 times per year
                   Biweekly payroll period   = 26 times per year
                   Semimonthly payroll period = 24 times per year
                   Monthly payroll period   = 12 times per year

STATE & FEDERAL

CONTRACTOR LAW QUIZ NO. 1

This is a timed quiz - allow 90 minutes.

1. If an employee is injured on the job primarily due to being intoxicated he shall be entitled to _____.

    A. 25% of his average weekly wage
    B. No compensation benefits
    C. 25% of regular compensation benefits
    D. Half of his weekly salary

2. If an injury results in disability for more than fourteen days, compensation shall be allowed from _____.

    A. Seven days after disability
    B. Commencement of the disability
    C. Three days after disability
    D. The fourteenth day after disability

3. In the case of total disability adjudged to be permanent, 2/3 of the pre-injury wages shall be paid to the employee _____.

    A. for 700 weeks
    B. for 350 weeks
    C. during the continuance of such total disability
    D. for one year

4. Willful refusal of the employee to use a safety appliance, resulting in injury, reduces the compensation by _____.

    A. 25%   B. 30%   C. 60%   D. 75%

5. The maximum compensation an employee can collect for a disability resulting from injury is subject to _____.

    A. $54.00 a week
    B. 60% of his average weekly salary
    C. 1/2 of his average weekly salary
    D. 100% of the statewide average weekly wage

6. Funeral expenses _____ will be paid for an employee whose death results from an accident or injury on the job.

    A. in total
    B. up to $1,000.00
    C. up to $2,500.00
    D. up to $100,000.00

State & Fed. Contractor Law Quiz No. 1

7. The employer, if not self insured, shall notify his carrier of
    an accident within 7 days after the employer has notice of the
    accident. Any employer who fails to so notify his insurance
    carrier shall be subject to a civil penalty _____.

    A. $500.00 fine and/or 6 months in jail
    B. $5.00 for each 30 days or fraction thereof
    C. not to exceed $100.00 for each failure or refusal
    D. $5.00 for each day after 10 days

8. Compensation for death payments to the spouse, if there are no
    children, shall be paid _____.

    A. at 50% of the average weekly wages
    B. at 66 2/3% of the average weekly wages
    C. up to $50,000 in workers' compensation
    E. up to $100,000 in workers' compensation

9. If any compensation payable under the terms of an award is not
    paid within 14 days after it becomes due there shall be added to
    such unpaid compensation an amount equal to _____.

    A. the compensation payable under the award
    B. 20% thereof
    C. 10% additional
    D. double the sum due

10. Any agreement between employee and employer to pay any portion
    of the premium for workers' compensation is _____.

    A. invalid
    B. permitted
    C. to meet with commission approval first
    D. valid if employer pays 60% of cost

11. All notices received by an employer who has secured workers'
    compensation shall be _____.

    A. posted in a conspicuous place available to the
       employees of his business
    B. forwarded to his insurance carrier
    C. sent by registered mail only
    D. copied and given to each employee

12. For the State, employers are required to maintain employment
    records for a period of _____.

    A. 1 year   B. 4 years   C. 350 weeks   D. 5 years

State & Fed. Contractor Law Quiz No. 1

13. The maximum weekly unemployment benefit an individual can receive is _____ .

    A. $74.00   C. $125.00
    C. 1/2 the average weekly wage   D. $150.00

14. Under the unemployment compensation law, the penalty for delinquent reports is _____ .

    A. $100.00 fine for each report
    B. $10.00 fine
    C. $5.00 for each 30 days or fraction thereof
    D. 20% of taxes due plus $5.00

15. According to the S.S.B.C., for a building exceeding 400 sq ft in area the minimum allowable depth for footings and foundations is _____ .

    A. 12 inches below grade   C. 16 inches below grade
    B. 14 inches below grade   D. 18 inches below grade

16. The Federal Unemployment tax rate is _____ .

    A. 3.5% of the first $7,000 earned
    B. 1% of total earnings
    C. 5.85% of total earnings
    D. 6.2% of the first $7,000 earned

17. The Federal Tax Deposit Form is _____ .

    A. Form W-2   C. Form 8109
    B. Form 940   D. Form 8109-A

18. The maximum amount of wages subject to Social Security Taxes for 1986 is _____ .

    A. $ 7,000.00   C. $39,600.00
    B. $ 35,700.00   D. $60,600.00

19. The Florida Construction Industry Licensing Board shall consist of _____ members.

    A. 10   B. 15   C. 17   D. 21

State & Fed. Contractor Law Quiz No. 1

20. A person not in privity with the owner as a prerequisite to perfecting a lien, shall serve a notice to owner _____.

   A. in 90 days from last date materials or services were provided
   B. within 60 days
   C. before commencing or not later than 45 days from commencing to furnish materials or services
   D. up to 1 year from last date service or materials provided

21. A recorded claim of lien unless contested shall not continue longer than _____.

   A. 90 days   B. 45 days   C. 1 year   D. 6 months

22. A notice of commencement shall be void and of no further effect if the improvement described in said notice is not actually commenced within _____.

   A. 90 days   B. 45 days   C. 30 days   D. 1 year

23. The contractor shall be responsible to the owner for _____.

   A. making sure all drawings and specs are in accordance with applicable laws, statutes, building codes and regulations
   B. furnishing all surveys
   C. for the acts and omission of all his employees or subcontractors performing any of the work under a contract within the Contractor
   D. securing and purchasing property insurance

24. The securing and paying for all permits, governmental fees and licenses necessary for the proper execution and completion of the work which is applicable at the time bids are received is the responsibility of the _____.

   A. Owner   B. Architect   C. Inspector   D. Contractor

25. Tube and coupler scaffolds over 125 feet in height must _____

   A. be constructed with heavy duty steel
   B. be constructed with light duty steel
   C. be designed by a qualified engineer, competent in this field
   D. have posts spaced no more than 6 feet apart by 10 feet along the length of the scaffold

State & Fed. Contractor Law Quiz No. 1

26. A trench is a narrow excavation made below the surface of the ground but the width of a trench is not greater than _____.

    A. 3 ft   B. 5 ft   C. 10 ft   D. 15 ft

A female employee is married to a nonresident alien. For withholding tax purposes she is considered _____.

    A. married with 2 withholding allowances
    B. married with 1 withholding allowance
    C. single with 2 withholding allowances
    D. single with 1 withholding allowance

A married person having 4 withholding allowances earns $220.00 per week and is paid weekly, what is the withholding and FICA amounts, respectively, to be withheld each week?

    A. $ 9.00 - $15.51   C. $ 9.00 - $15.73
    B. $11.00 - $15.51   D. $11.00 - $15.73

29. When the contract for improving real property is made with a husband or wife who is not separated and living apart from his or her spouse and the property is owned by the other or by both the spouse who contracts shall be deemed to be the agent of the other to the extent of subjecting the right, title, or interest of the other in said property to liens under Part I of Chapter 713 (Fla. Lien Law) unless such other shall within _____ after learning of such contract, give the contractor and record in the clerk's office, notice of his or her objection thereto.

    A. 3 days   B. 5 days   C. 10 days   D. 15 days

30. If you are injured on the job or have a work-related illness, you have a right to _____.

    A. medical treatment after 10 days
    B. medical treatment when approved by employer
    C. medical treatment when approved in writing
    D. immediate medical treatment

ANSWER SHEET

STATE AND FEDERAL LAW QUIZ NO. 1

```
 1.  B -  A/Z, S
 2.  B -  A/Z, W
 3.  C -  Workers' Comp. p 9
 4.  A -  A/Z, S
 5.  D -  A/Z, Y
 6.  C -  Workers' Comp., p 4
 7.  C -  A/Z, P
 8.  D -  Workers' Comp., p 4
 9.  C -  A/Z, X
10.  A -  Workers' Comp., p 2
11.  A -  A/Z, F
12.  D -  Unemployment, p 31
13.  D -  Unemployment, p 36
14.  C -  Unemployment, p 22
15.  A -  SSBC, section 1302.1
16.  D -  Circular E, Front Cover
17.  C -  Circular E, p 8
18.  D -  Circular E, Front Cover
19.  C -  Chapt. 489 - 489.107 (2)(a)
20.  C -  Lien Law, 713.06 (2)(a)
21.  C -  Lien Law, 713.22 (1)
22.  C -  Lien Law, 713.13 (2)
23.  C -  AIA, 4.3.2
24.  D -  AIA, 4.7.1
25.  C -  OSHA, 1926.451 (c) (4)
26.  D -  OSHA, 1926.653 (N)
27.  D -  Circular E, p 7
28.  D -  Circular E, p 24 and 7.15% of $220.00
29.  C -  Lien Law, 713.12
30.  D -  Workers' Comp., p 6
```

INDEX